Transforming China's Public Services

Keyong Dong • Na Wei

Transforming China's Public Services

A Plan for 2030

palgrave
macmillan

Keyong Dong
Renmin University of China
Beijing, China

Na Wei
Renmin University of China
Beijing, China

ISBN 978-981-99-3942-8 ISBN 978-981-99-3943-5 (eBook)
https://doi.org/10.1007/978-981-99-3943-5

This Palgrave Macmillan imprint is published by the registered company Springer Nature Singapore Pte Ltd.
The registered company address is: 152 Beach Road, #21-01/04 Gateway East, Singapore 189721, Singapore

Preface

This book is the outcome of the cooperation of a dozen of academics at Renmin University of China (RUC) under "Modernization of Basic Public Services: The Way Forward to 2030," a project entrusted by the Department of Social Development of the National Development and Reform Commission of the People's Republic of China, and undertaken by the RUC School of Public Administration, several years ago. The project was designed to envisage China's basic public modernization from 2010 to 2030 by studying and analyzing the status quo of and problems in basic public services and to explore the way toward basic public services with Chinese characteristics. The members of the research team conducted intensive research and analysis on the basic content of the modernization of China's basic public services from 2010 to 2030 and approaches to realize it, including research on theories and international experience and the forecast of the overall economic and social trend from 2010 to 2030, and concluded the project with a report. After that, the researchers have followed up on the topic, extended and optimized the findings in their respective research areas, and presented the results to readers in this book with the support of Renmin University of China Press.

In every country, public service provision is one of the basic functions of the government. However, what is the scope of public services, and what are basic public services? What is the objective of public services or basic public services? How can basic public services be provided effectively? And some more concerning questions: How are public services in China now, and what will it be like in 2020 and 2030? The book begins with some brief answers to the basic questions on public services.

First, what are public services? Public services, in a narrow sense, are primarily to meet basic livelihood needs in the areas of education, employment, social security, medical and health care, housing security, culture, and sports. In a broader sense, the term also includes public services in areas that are closely related to people's living, including transportation, communications, public facilities, and environmental protection, and those in areas for safeguarding people's security, including public security, consumer security, and national defense security. As space is limited, the book focuses on public services in a narrow sense for basic livelihood needs.

Second, can the market meet the needs for public services? In other words, can the market ensure that everyone's needs are met fairly and efficiently? The answer is negative. For some products and services, especially public goods, the market is either unwilling or unable to provide it effectively, which is the economically so-called market failure. In that case, it should be the government that performs such functions.

Third, what are basic public services? Basic public services are led by the government, secure the basic needs of all citizens for survival and development, and are compatible with the current economic and social development. Basic public service is a concept raised by China based on its national conditions and development stage. It covers every stage of a citizen from birth to death, and basic services in the areas of education, labor, employment and entrepreneurship, social insurance, medical and health care, social services, housing, culture, and sports.

Fourth, what are the characteristics of basic public services? There are three of them: basic, adaptive, and equal. Basic means that basic public services meet basic livelihood needs; citizens have the right to receive basic public services; and it is the government's function to provide basic public services. Adaptive means that the service quality of basic public services should match social and economic development, not lagging or exceeding. The means and methods of providing services should be compatible with infrastructure construction and scientific and technological progress. Equal means that most citizens can get equal access to basic public services. Its core is to promote equal opportunities, and the focus is to ensure the people's access to basic public services, instead of the simple egalitarianism. It is a basic right of citizens to receive basic public services. It is a crucial function of the government to ensure that everyone receives basic public services.

Fifth, why are basic public services so important? From the characteristics of basic public services, it can be seen that they are conducive to fundamentally fulfilling citizen' rights, realizing a sense of gain among the people, and enhancing national cohesion. The provision of basic public services facilitates the transformation of the economic development pattern and promotes the sustainable development of the economy and the expansion of domestic demand. The equal provision of basic public services promotes social stability and solidarity and sets the foundation for the long-term stability of the country. The provision of basic public services also helps narrow the income gap and improve people's livelihood. More importantly, the effective provision of basic public services is critical to the modernization of the country's governance system and governance capacity.

Sixth, who should be the provider of basic public services? Although it is the government's function to provide basic public services, it does not mean that all basic public services must be provided by the government. From the lessons of many countries in the world, public services directly provided by the government often account for only a fraction. In China, most public services are provided by public institutions, such as schools and hospitals. In other countries, the government may purchase public services or essential products and services from non-profit organizations and enterprises.

The book is the fruit of collective wisdom. Professor Dong Keyong, the leading editor, is the chief planner and contributor to the main ideas. Other participants in the project are

Wei Na (professor at the School of Public Administration, RUC)
Zhou Guangli (professor at the School of Education, RUC)
Li Wenzhao (professor at the School of Public Administration, RUC)
Liu Peng (professor at the School of Public Administration, RUC)
Tang Jie (associate professor at the School of Public Administration, RUC)
Ouyang Wei (lecturer at the School of Public Administration, RUC)
Zhou Xiang (associate professor at the School of Education, RUC)
Yang Shenghui (postdoctoral researcher at the School of Public Administration, RUC)
Dong Pengtao (doctoral candidate at the School of Public Administration, RUC)
Yang Yuefeng (doctoral candidate at the School of Public Administration, RUC)
Liu Hailan (doctoral candidate at the School of Education, RUC)

Liu Changqian (doctoral candidate at the School of Public Administration, RUC)
Chen Xi (graduate student at the School of Public Administration, RUC)

The book was divided in parts for compilation. Specifically, Chap. 1 was compiled by Li Wenzhao; Chap. 2 by Liu Peng; Chap. 3 by Zhou Guangli and Liu Hailan; Chap. 4 by Ouyang Wei; Chap. 5 by Dong Pengtao; Chap. 6 by Wei Na and Yang Yuefeng; Chap. 7 by Tang Jie and Yang Shenghui; Chap. 8 by Wei Na, Liu Changqian, and Chen Xi.

We hope that the publication of this book will attract more attention from experts, scholars, government officials, and the general public to this issue. Hopefully, it will intensify people's comprehension and understanding of the modernization of basic public services, better build consensus on this cause, and accelerate the pace of its realization.

Beijing, China Keyong Dong

CONTENTS

List of Figures

List of Tables

CHAPTER 1

Modernization of Basic Public Services: Theoretical Logics and Path Selection

1.1 Introduction

When developing the blueprint and selecting a path for the modernization of China's basic public services by 2030, we need to think about four basic questions. First, where are we? It is mainly about the status quo of and problems in the modernization of China's basic public services. Second, where are we going? The answer should describe the goal and the blueprint for the modernization of China's basic public services by 2020 and 2030. Third, what should we do? To be specific, in the course of achieving the blueprint and the goal, how should we overcome difficulties and find solutions? What is the priority in resource allocation? And how to find action tactics? Fourth, why should we act this way? It concerns the theoretical logic, historical experience, international experience, and expert judgment of the above actions, focusing on the question that why these actions are effective and reasonable. The logical relation of these four questions is shown in Fig. 1.1.

To answer these questions, we must first identify the nature of the modernization of basic public services, especially from the perspective of China in 2030. The importance of the modernization of basic public services can be summarized in four aspects: politics, economy, society, and state governance. The modernization of basic public services is conducive to (1) enhancing legitimacy, (2) fundamentally implementing civil rights

K. Dong, N. Wei, *Transforming China's Public Services*,
https://doi.org/10.1007/978-981-99-3943-5_1

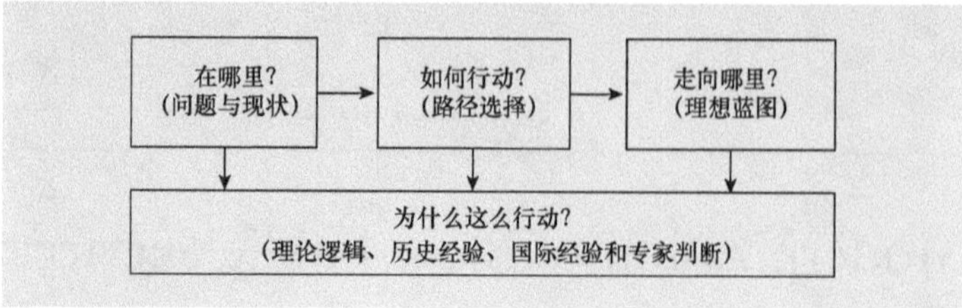

在哪里？（问题与现状）	Where are we? (problems and status quo)
如何行动？（路径选择）	What should we do? (path selection)
走向哪里？（理想蓝图）	Where are we going? (the blueprint)
为什么这么行动？（理论逻辑、历史经验、国际经验和专家判断）	What should we do this way? (theoretical logic, historical experience, international experience, and expert judgment)

Fig. 1.1 Research framework of the modernization of China's basic public services

and realizing citizen treatment in the community, (3) sustainable economic growth and the provision and guarantee of social infrastructure, and (4) maintaining social stability and unity and achieving long-term peace in the country.

1.2 Comprehending the Modernization of Basic Public Services: A Conceptual Analysis Framework

Modernization is a key analytical framework for understanding the societal changes and transformations of the world and China. By understanding modernization, we can better understand the connotation of the modernization of basic public services. Modernization originated in the ideologies of the Reformation, the Renaissance, and the Enlightenment in Europe. These movements justified secular goals and personal choices and laid the groundwork for the subsequent industrial, social, institutional, and technological revolutions. Modernization embraces at least rationalization, commercialization, industrialization, urbanization, impersonalization, and separation of public and private. It entails all-round changes in technology, ideas, systems, culture, lifestyle, social communication, production techniques, and living forms. It is a process in which a new order replaces the old one, a process of not only active adaptation but also passive response.

From the perspective of modernization, we can gain more insights in the modernization of basic public services. It is a crucial and even essential component of modernization. Basic public services being modernized means that the provision and production of public services must reflect some of the basic characteristics of modernization, such as rationality, efficiency, professionalization, industrialization, and urbanization. Rationalization is one of the paramount characteristics of modernization. Its essence is to find the optimal means to achieve some goal. It requires that public service provision and production conform to the principles of rationality. Modernization emphasizes the division of labor, professionalization, and efficiency. It entails professional provision and production of public services, that is, public services are provided by professionals and professional organizations, instead of self-sufficiency. Public service organizations must be independent from the government and the private sector. They must be independent corporations and take responsibility for their actions. Corresponding to professionalization, the modernization of basic public services requires the bureaucratization of public service organizations, which coordinates, motivates, and integrates through clear regulations, ranks, and salaries, among other means. Emerged in the context of industrialization and urbanization, modernization requires public service provision and production to reflect this attribute. Urbanization emphasizes the difference between urban and rural public services. The fundamental characteristics of cities are spatial agglomeration and scale expansion, and large-scale industrial production raises new demands for public services. Urbanization and industrialization will have a profound impact on public service arrangements, provision, and production. Modernization is always the forerunner of technology. As technology promotes industrialization and urbanization, it will inevitably have a profound impact on public service organizations and production. Modernization also means the separation of public and private spaces. The protection of public space is vital to public services. It is primarily manifested in the socialization of some areas and affairs traditionally held by individuals, such as education, aged care, and medical care. It is also the inevitable result of a risk society and increasing uncertainty, that is, the public bears public risks.

It can be concluded that the modernization of basic public services features at least rationalization, professionalization, bureaucratization, urbanization, intelligence, and publicization (see Fig. 1.2). Rationalization

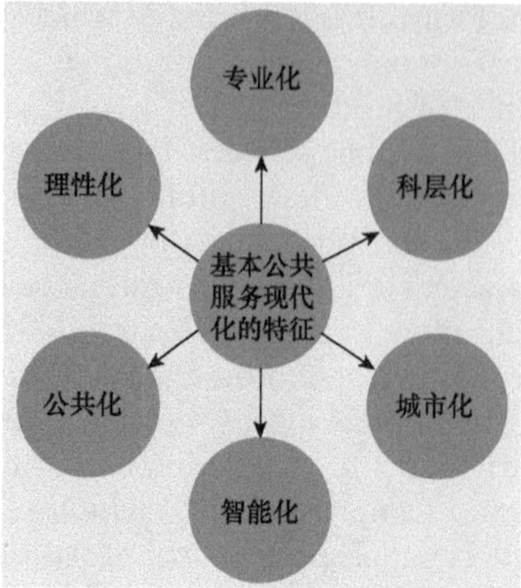

基本公共服务现代化的特征	Features of the modernization of basic public services
专业化	Professionalization
科层化	Bureaucratization
城市化	Urbanization
智能化	Intelligence
公共化	Publicization
理性化	Rationalization

Fig. 1.2 Features of the modernization of basic public services

means that basic public service organizations need to consider benefits and costs. Professionalization means that basic public service organizations need to consider the division of labor and efficiency. Bureaucratization is the organizational basis for the above two features. It requires proceduralization, standardization, and impersonalization, and links everyone's remuneration with their ability. Urbanization means that basic public service organizations need to consider space and scale. Intelligence means that basic public service organizations need to consider scientific and technical issues and apply artificial intelligence as much as possible in addressing public service supply and production, for example, the revoluton means Massive open online courses revolution (MOOC) revolution in education and the machine diagnosis in medical care. Publicization means that basic public service organizations need to show fairness and justice. They must be open, fair, and just, and treat everyone in one category equally without discrimination.

Any basic public service can be analyzed from the perspectives of actors and processes, that is, which actors are involved in the service and what processes they will go through. Moreover, service attributes, institutions, technology, and cultural attributes will have a significant impact on the actors and the processes.

Following the above-mentioned two threads, actor and process, the conceptual analysis framework for the modernization of basic public services focuses on public services, government, citizens, and organizational model, and discusses the inherent logical relationship between these four elements (see Fig. 1.3). It concerns three main questions: First, what falls in the scope of public services? Second, what kind of organizational model should be adopted? Third, what is public service governance? The answers to these questions determine the main functions of the government and citizens. Different countries and societies have different answers to these questions, thus forming different public service systems. The discussion on the first question is directly related to public service provision. That on the second question is directly related to public service production. The third question concerns the linking between provision and production, and the respective oversight over the two processes. In general, the government and citizens are directly involved in the provision process, determining the financing, content, and distribution of use of public services. For the production process, the government and the public mainly play the roles as supervisor and cooperative producer. Accordingly, we can divide public services into processes and segments and discuss the roles and functions of the government, citizens, society, market, and organizations in them. The provision, production, and governance all need the support of institutions, technology, and culture, so that the entire process can operate in an open, fair, efficient, and orderly manner.

Public service provision should follow the principles of fiscal equilibrium, democracy, social justice, and sustainability. The principle of fiscal equilibrium was first put forward by Mancur Olson in the discussion on the government structure theory. The core idea of this principle is that "beneficiary pays," which applies to financing and payment in public service provision. According to this principle, a public service within the purview of the central government shall be paid by the central government, and that within the purview of the local government shall be paid by the

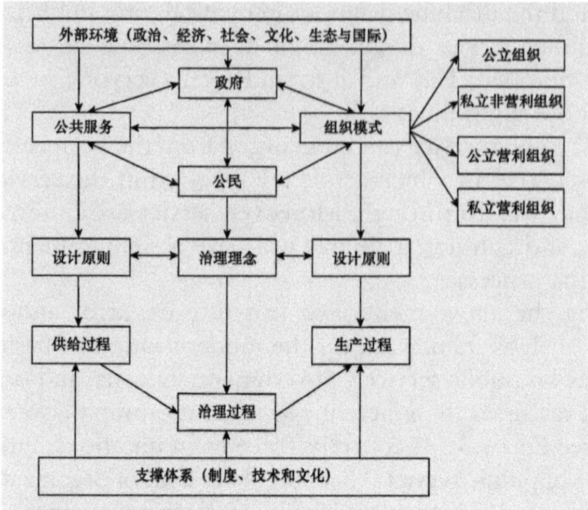

外部环境（政治、经济、社会、文化、生态和国际） External environment (political, economic, social, cultural, ecological, and international)			公立组织 Public organizations
	政府 Government		私立非营利组织 Private NGOs
公共服务 Public services		组织模式 Organizational model	公立营利组织 Public profit organizations
	公民 Citizens		私立营利组织 Private profit organizations
设计原则 Design principles	治理理念 Governance ideas	设计原则 Design principles	
供给过程 Provision process		生产过程 Production process	
	治理过程		
	Governance process		
支撑体系（制度、技术和文化） Support systems (institutions, technology, and culture)			

Fig. 1.3 A conceptual analysis framework for the modernization of basic public services

local government. All these payments come from taxes, which means that if a region needs more public services, taxpayers in the region have to pay more in taxes. If a public service provided by a local government is paid by the central government, it will lead to oversupply or undersupply. The principle of democracy emphasizes that under the principle of fiscal equilibrium, what kind of public services a region shall provide and to what extent it shall provide them should be decided by the citizens in the region through democratic procedures. Public service provision in the region shall reflect the preferences and willingness to pay of citizens in the region. The principle of democracy means that the local government should provide more choices in public service provision and ensure citizens have the practical right to choose. The principle of social justice emphasizes that the government, especially the central government, should provide some basic public services that are conducive to ensuring the basic fairness and justice and maintaining the basic stability and social equality. For these basic public services, whether within the direct responsibility of the central government or the joint responsibility of the central government and the local government, they are all "hard constraints" and must be provided. The principle of sustainability emphasizes that the provision of basic public services should be sustainable and match the level of economic and social development of the country or region. Over-providing public services may lead to unsustainability.

Public service production should follow the principles of co-production, efficiency, competition, and diversity. The principle of co-production was summarized by Elinor Ostrom in the research on policing, which highlighted the importance of civic engagement to public security. The principle requires citizens to participate in public service production and work with police organizations to produce public security. Besides policing, citizens' conscious participation is crucial for better results in education, medical care, and aged care. The principle of efficiency was advocated by Herbert Simon in organizational studies. Efficiency means to get the most at the lowest cost or, specifically, to maximize the benefit at a given cost or to minimize the cost for a given benefit. The principle of efficiency requires public service organizations to act at the maximum efficiency and find the most efficient way to provide public services. The principle of competition is advocated mostly by economists. They regard competition as a process of discovery, which can achieve the optimal allocation of resources. The New Public Management theory introduced the principle of competition into the public domain, emphasizing that competition between public service organizations can improve the efficiency of public service provision. The principle of diversity was first advocated by biologists, who emphasize the diversity of life. The application of the principle of diversity in public service organizations means that public service organizations shall be of various types and sizes and the organizational model of public services shall match the attributes of public services. Capital- and labor-intensive public services differ in the organizational model. Therefore, public service organizations shall be diverse, as diversity of public service organizations encourages competition and improves efficiency.

Public service governance should follow the principles of working together, regulation, responsibility, and trust. The principle of working together is fundamental to governance. It means that public service provision and production are organically unified, and that the government, citizens, and public service organizations act together to produce and provide public services. Different ways to combine the production and provision of different public services will bring different institutional arrangements for public services and thus different actor roles and functions. The principle of regulation is an important facilitator for governance. It means that the government plays an important role in governance, or the government shall play a role in meta-governance. The regulation here refers to that the government regulates public service organizations on whether they produce quality public services in accordance with the choices and requirements in public service provision. The principle of responsibility is that the government, citizens, and public service organizations should assume corresponding responsibilities. Public service organizations must keep their mission in mind and produce high-quality public services. The government should act as regulator, respond to citizens' needs, and hold public service organizations accountable. Individual citizens should assume the roles and responsibilities as co-producer and supervisor. According to the findings in *Making Services Work for Poor People*, the World Bank's World Development Report 2004, citizens can hold public service organizations accountable via the "short route," that is, to directly voice and exercise oversight over public service organizations. They can also do that via the "long route," that is, by reporting problems in public service production to the government, which will oversee public service organizations. The principle of trust emphasizes building trust among the government, citizens, and public service organizations. Only with trust can the three parties act together, and only in this way can they take responsibility more actively and proactively and solve problems in the provision and production of public services, such as information asymmetry and conflict and inconsistency of interests.

From the analysis of the provision, production, and governance of public services, we can see more clearly that in public service provision, the government and citizens play an important role, who are similar to collective consumers. Between the two, the government plays a key role, primarily because it is entrusted by citizens to make choices about the scale and scope of public services and to plan the use of public services and manage their availability, that is, who have the access to public services and on what terms they have such access. Citizens exercise oversight of the government in political procedures. In public service production, public service organizations and citizens play an important role, who are similar to collective producers. Between the two, public service organizations, such as public hospitals, schools, and aged-care institutions, play a key role, primarily because they are the direct producers of public services. Citizens only act as co-producers. In public service governance, the government, citizens, and public service organizations all play an important role. Citizens are the key, mainly because they are the final consumers of public services and important evaluators for the provision and production of public services. In addition, citizens hold the government and public service organizations accountable through oversight. Table 1.1 analyzes the public service system in terms of processes, principles, and actors.

In public service provision, production, and governance, public services' attributes, institutions, technology, and culture have a profound

Table 1.1 Processes, principles, and actors in the public service system

Process	*Principle*	*Actor*
Provision	Fiscal equilibrium Democracy Social justice Sustainability	The government and citizens
Production	Co-production Efficiency Competition Diversity	Public service organizations and citizens
Governance	Working together Regulation Responsibility Trust	The government, citizens, and public service organizations

impact on the entire process. Different types of public services have different provision, production, and governance processes. Problems in public service provision, production, and governance can be addressed with institutional, technological, and cultural changes. Institutions emphasize the regulation of the relationships between various subjects and public service processes. Both the government and public service organizations can harness institutions as an important tool to solve the problems in the public service system. Technology emphasizes the introduction of the most advanced technology, especially information technology, to solve the problems in public service processes, making the public service system intelligent. Culture emphasizes finding a culture that is compatible with the new public service system. For example, when a public service organization transforms from a government subsidiary to an independent corporation, it also needs to change its culture, that is, from a bureaucratic culture to a service-oriented culture.

1.3 Evolution of and Problems in the Modernization of Basic Public Services

In the discussion on path selection for the modernization of China's basic public services, it is vital to review the evolution of this cause and analyze its current problems and challenges, which is a process of diagnosis. The modernization of China's basic public services is closely related to China's economic, political, and social system reforms. On the one hand, it is affected by the "great transformation" of the entire order; on the other hand, it may promote or hinder the "great transformation." This is the interaction between the environment and the modernization of China's basic public services mentioned in the previous part.

In the "great transformation" of the order, the reform of the economic system has a profound impact on the modernization of China's basic public services. The reform of the economic system is the pacemaker for the "great transformation" of the entire order, leading reforms in other areas. Naturally, the basic public service system is affected by the transformation of the economic order. The economic order and the modernization of basic public services are isomorphic in many aspects. It means that the process of economic transformation from a planned economy to a market economy calls for the transformation of public services. The modernization of basic public services is integral to a modern market economy. The "great transformation" of the economic order also drives government reform. The government needs to adjust its functions and finally make public service provision one of its key functions, which will also promote reform and innovations in the modernization of basic public services. Figure 1.4 depicts the inherent logical relationship between order transformation and the modernization of basic public services.

According to the logic of Fig. 1.4, we can divide the evolution of the modernization of China's basic public service into four stages from the perspective of the "great transformations" of China's economic, political, and social orders. They are the planning period, the transition period, the institution period, and the modernization period. The division is mainly based on the transformation of the economic order. The planning period is mainly about the composition of the public service system in the planned economy period. The transition period is mainly about the reform of the public service system in China's transition from a planned economy to a market economy. The institution period is mainly about the deepened

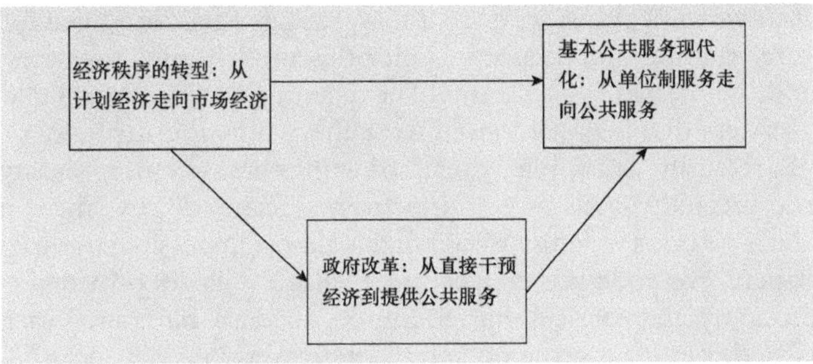

经济秩序的转型：从计划经济走向市场经济	Transformation of the economic order: From planned economy to market economy
政府改革：从直接干预经济到提供公共服务	Government reform: From direct intervention in economy to public service provision
基本公共服务现代化：从单位制服务走向公共服务	Modernization of basic public services: From entity-based services to public services

Fig. 1.4 The internal logical relationship between order transformation and the modernization of basic public services

reform of China's basic public services after the establishment of the market economy. The modernization period is mainly about the modernization of basic public services against the modernization of China's economic and social development. These four periods correspond to the four key periods of the "great transformations" of China's economic and social orders. The planning period represents the "old orders" of the economy and society. The transition period represents the "great transformation" of the economic and social orders. The institution period represents the process of establishing "new orders" of the economy and society. The modernization period represents the improvement and maturity of the new economic and social orders. Figures 1.5 summarizes the basic characteristics of the economy, society, and public service systems in the four periods.

Strictly speaking, there were no real "public services" in the planning period. The whole society was organized under the planned economy model and divided into two communities, rural society and urban society. The two communities differed in the public service provision mode. In urban society, everyone was attached to a *danwei*, or entity. It could be an enterprise, the government, or a public agency. Entities provided people with what we call "public services" today, such as education, medical care, aged care, culture, and security. Under this model, public services were nonpublic but nested in each entity. The quantity, quality, type, and level of public services that people received were affected by the attributes of their entity. Each entity, as a whole, interacted with other systems in society. The existence of individuals was distinctively "featured" by their entity. Individuals without an entity would find it almost impossible to survive. In rural society, everyone was attached to a village. Villages provided people with economy and services and education, medical care, aged care, and cultural activities. "To grow up and die wherever you were born" vividly describes this traditional and static social form. Public service provision could never be fair in the planning period. There was inequity between urban and rural areas and among cities. The more capable and powerful an entity was, the more economic power and services it could obtain. An entity's access to resources mainly depended on its relationship with the government and the level of the government itself. The government, especially the central government, was the ultimate distributor of all the resources of the whole society. They included both economic resources and public service resources. From the perspective of actor and process of public service provision, the provision, production, and governance of public services were centralized and provided by entities. Of course, there were

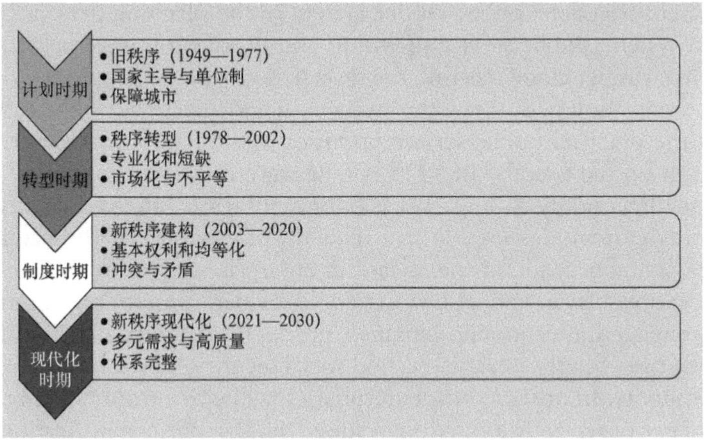

The Planning Period	Old order (1949—1977)
	State-leading and entity-based
	Securing cities
The Transition Period	Order transition (1978—2002)
	Professionalization and shortage
	Marketization and inequality
The Institution Period	Construction of the new order (2003—2020)
	Basic rights and interests and equalization
	Conflicts and contradictions
The Modernization Period	Modernization of the new order (2021—2030)
	Diversified demands and high quality
	A complete system

Fig. 1.5 Evolution of the modernization of China's basic public services

also a very small number of real "public" services for everyone, provided that they were either impossible or uneconomic to be provided by entities, such as large hospitals and universities. (Even a university was exclusive to a certain industry and not open to the public.) It can be seen that state leadership, the entity-based system, and securing cities are the three basic characteristics of the public service system in the planning period. Under such a system, public service provision and production were neither efficient nor fair, let alone meeting the diverse needs of the people.

The transition period saw the awakening and advent of public services, when the practical public service provision, production, and governance were put on the agenda. In 1978, China started the transformation from a planned economy to a market economy. In 1992, the socialist market economy system was established. The transformation of the economic order inevitably required the release of enterprises from the double roles of economy player and public service provider. Enterprises should be solely engaged in economic activities, production, and profit. Education, medical care, healthcare, culture, and social security should not fall in their responsibility. In other words, enterprises shall be free from "serving society." This order transformation resulted in the professionalization and shortage of public services. Professionalization means that public service organizations shall provide public services to indeterminate objects rather than just being affiliated with an agency or organization. Shortage means that public services require more investment. With insufficient fiscal funds, public service provision and production were always in short supply against the public's demand, especially high-quality public services. We can see that the transformation of the economic and social orders required the transformation of the public service system, from decentralized to public provision, production, and governance, to make such services "public." However, the exemption for the government from such responsibilities created a contradiction between professionalization and shortage. On the one hand, professionalization demands necessary resources to meet professional standards. Professionalization, which entails independence and autonomy, must follow the logic of profession and knowledge. On the other hand, with insufficient fiscal investment from the government, in order to maintain public service provision and mobilize public service producers, it was necessary to introduce "market logic" into public service provision, production, and governance. It demanded consumers' payment to fund public services.

The contradiction between professionalization and shortage constrained further professionalization and prevented public services from serving the public. At that time, winners and losers in the market economy were reflected in the consumption of public services. Only the winners could receive high-quality and scarce public services, while the losers could not, which escalated inequality. Many sociological researchers use the term of "class reproduction." A significant factor leading to "class reproduction" is the difference in the access of different classes to educational resources. The government concentrated limited public service resources to a few areas and certain groups. The allocation of resources by government privilege further exacerbated the inequality created by the market economy. In an article titled "Shanghai should prevent the separation of the rich and the poor as soon as possible," Li Huiping discussed spatial social stratification, or differences in the spatial allocation of public services in the city, and its impact on the economy and society. His key argument is that it leads to the segregation of the rich and the poor in the city to allocate more public service resources to richer regions while less to the poorer. Compared with in the planning period, public service organizations had more autonomy during the transition period, but public service provision, production, and governance were still divided. Public service resources were accessible to only those related. It can be seen that professionalization, shortage, marketization, and inequality basically characterized the basic public service system in that period. Under that system, there was limited competition between public service organizations, but it was not sufficient. Moreover, inequality in the access to public services was intensified.

The institution period is a period when public services manifest publicness. It is a period of rethinking of the publicness of public services and the equal rights of every citizen. It started in about 2003 and extended to around 2020. Obviously, the SARS crisis in 2003 was the focal event of the period and the "fuse" that started a new order. It was an opportunity for society as a whole to reexamine the equity of public services. Public service equalization has become a key issue in the construction and design of the basic public service system in this period. Its core is to expand the coverage of public service access, so that more people can receive public services. The concept of "basic public service" is a value proposition. It has always been associated with civil rights, in an attempt to continuously expand the range of recipients of basic public services. However, as citizens are being aware of "basic public services" and their demand for basic public services is expanding, this will inevitably lead to conflicts and contradictions in public services that are provided limitedly, produced with low quality, and governed defectively. From the perspective of equality, there are still the three major challenges to the modernization of basic public services: regional differences, urban-rural differences, and class differences within the city. It can be expected that as the equality issue is addressed, the quality of public services will be brought back to the agenda, that is, how to improve the efficiency and quality of public service provision, production, and governance, so that more people have access to high-quality and efficient basic public services. It is called the "institution period" mainly because it demands a set of institutional arrangements for the provision, production, and governance of basic public services to standardize the rights and obligations of governments at different levels, citizens, and public service organizations, to catalyze a modern basic public service system. With institutional construction, the basic public service system will form a new order, lifting the provision, production, and governance of basic public services to a new level. Another key feature of this period is the gradual inclusion of farmers in access to basic public services. For example, the new rural cooperative medical care system has been implemented in rural areas to provide farmers with medical insurance services. A basic pension system has been implemented in rural areas, granting pensions to elderly people, although the amount and scope of such pensions are very limited. Compulsory education has been further popularized in rural areas. To conclude, equalization, citizenship, conflicts, and contradictions are the main features of the institution period. It is also a critical period for the modernization of basic public services. Without a sound institutional foundation for the modernization of basic public services, it will be difficult to establish a new order of the provision, production, and governance of basic public services.

The modernization period is the guarantee for the modernization of basic public services, which will be a focus of China in the reform and innovation of basic public services from 2021 to 2030. The modernization of basic public services is only a vision now, but it can be envisioned that in addition to solving the problems and challenges in the previous three periods, it also needs to meet diverse demands, provide high-quality services, and form a complete system that is efficient, fair, and sustainable. The goals of this period and how to achieve them will be the focus of the next section.

1.4 Path Selection for the Modernization of Basic Public Services

In the first part, we analyze the significance to modernize basic public services. In the second part, we develop a theoretical model for the modernization of basic public services. In the third part, we analyze the evolution of the modernization of basic public services and problems in it.

These three parts give clearer answers to "where we are," "where we are going," and "why we should do that" for China's basic public services. Now comes the last part, which is to discuss the path from "where we are" to "where we are going," and its strategy and tactics. Based on the theoretical model in the second part and the problem analysis in the third part, it is reasonable for China to take the following paths for the modernization of basic public services:

1.4.1 Developing a New Governance Concept for the Modernization of Basic Public Services

According to the analysis in the third part, the main problems facing the modernization of China's basic public services lie in equity (equalization), professionalization (quality), diversification (demand), and efficiency (economy). To solve these problems, it is necessary to change the mindset for basic public service governance, replace the "old order" of basic public services with a "new order," and establish a new basic public service system.

The new governance concept should contain some design principles for public service provision, production, and governance, including fiscal equilibrium, democracy, social justice, sustainability, co-production, efficiency, competition, diversity, working together, regulation, responsibility, and trust. These principles shall serve to meet diverse needs, provide high-quality services, and form an efficient, fair, and sustainable basic public service system.

1.4.2 Setting the Scope, Scale, and Eligibility of Basic Public Services

In the modernization of basic public services, the first thing is to set the scope, scale, and eligibility of basic public services. We need to answer "who can get a certain kind of basic public services" and "how many basic public services are available." Ideally, there will be a list of basic public services, which may be revised over time and as citizens' needs change. However, in a given period, there should be one list as the basis for citizens to obtain basic public services and to oversee the government and public service organizations.

Due to regional differences, different regions' lists of basic public services must differ from each other. In general, there should be two lists. One is the list of basic, fundamental, and essential basic public services provided by the central government. All Chinese citizens will have access to these public services, which is their fundamental right. It is the minimum standard for local governments' lists of basic public services. The other is the list of differentiated, regional, and locally featured basic public services provided by the local government. Despite regional differences, local governments shall maintain a degree of openness in access to basic public services. Today, many local governments link access to basic public services to the household registration system. In the future, the household registration system should be phased out, and an open, fair, and just access system should be established.

1.4.3 Establishing a Diversified Service Provision System Based on Citizens' Needs and Choices

Basic public services, nonbasic public services, and services form a spectrum. Basic public services are essential in the modern service system but can only meet the fundamental needs of citizens. For nonbasic needs, citizens shall have the right to choose from different options. To meet citizens' nonbasic needs, it is necessary to establish a diversified service provision system, so that basic public services and nonbasic public services can develop in coordination.

Driven by the "externality of knowledge," the development of nonbasic public services and services boosts basic public services. The payments of citizens who are able and willing to pay for nonbasic public services or services contribute to the development of the entire system. For example, the development of private for-profit hospitals facilitates or provides a reference system for the development of public hospitals.

1.4.4 Establishing Efficient, Professional, People-Centric, and Competitive Modern Public Service Organizations

In modern society, almost all basic public services, whether purchased by the government or provided by public organizations, are produced by specialized public service organizations with professionals and specialized facilities. For these services, production and consumption concur. Doctors producing medical services is the process where patients purchase medical services. Teachers providing educational services is the process where students purchase educational services. The simultaneous occurrence makes regulation difficult. The biggest challenge faced by public service organizations is to provide good incentives and proper constraints for professionals and technicians.

According to the characteristics of modern organizations, modern public service organizations should be efficient, professional, people-centric, and competitive. Efficient means maximizing benefits with minimal cost and saving time for service recipients. Professional is the guarantee for quality and one of the fundamental characteristics of public service organizations. It means professionals providing professional services with their expertise, know-how, and experience. People-centric is determined by the characteristics of services. Basic public services ultimately need to meet the needs of service recipients, who, however, significantly differ from each other. It is one of the major challenges faced by public service organizations to balance standardization and differentiation and to balance unification and diversification. Competitive means the need to promote competition among public service organizations. Competition can improve efficiency and provide a platform for the flow of professionals.

1.4.5 Establishing a Professional, Rule-of-Law, Responsible, and Responsive Regulation System for Modern Public Services

A key feature of the modernization of basic public services is separating public service provision, production, and governance. As a crucial part of governance, regulation plays a critical role in improving the quality of basic public services and addressing the information asymmetry in the production of basic public services.

Since the objects under oversight are professional, the modern public service regulation system must first be professional. Sufficient expertise is the prerequisite for effective regulation of modern public services. Also, the modern public service regulation system must be based on rule of law. It requires the government to improve the institutional system for basic public services, so that the provision, production, and governance of basic public services have rules to follow. In addition, public service regulators must have a sense of responsibility. It can give better play to the initiative of regulators and prevent them from being controlled by the objects under oversight. Finally, the regulation system must be able to respond to demands, and promptly form a long-acting solution mechanism according to the characteristics of the demands.

1.4.6 *Establishing a Public Service System Based on Partnerships Between Citizens, the Government, and Public Service Organizations*

The modernization of basic public services requires the cooperation, joint action, and mutual trust of actors, which needs partnerships in the basic public service system. They can be partnerships between citizens and the government in public service provision, between citizens and public service organizations or between public service organizations in public service production, or between the government and public service organizations in the governance process.

To build partnerships, it is necessary to create a platform for the parties to interact. Only when actors resolve their shared problems through conversations, consultations, negotiations, communications, and listening can they develop situational awareness and have an understanding of the behaviors of others. This consensus is the foundation for common action and common belief.

1.4.7 *Modernizing Basic Public Services Based on Modern Systems, Technology, and Culture*

Systems are an important tool for human beings to solve the problems they face. They provide certainty, save transaction costs, and increase efficiency. The provision, production, and governance of basic public services need open, fair, and just systems, so that the basic public service system is subject to institutional constraints. Basic public services' financing, decision-making, distribution, responsibility, production, cooperation, and regulation in the provision, production, and governance all demand the normative systems. For example, for public service organizations, we need to establish personnel systems, information disclosure systems, and accountability systems. These systems must be open, enforceable, and effective. They must be made public and known to all stakeholders, as open systems ensure social oversight. These systems must be enforceable, so that they can be truly implemented, otherwise they are just a piece of paper. In addition, it is necessary to evaluate the effectiveness of the systems and set the conditions for the revision of them.

Information technology is changing human society. The modernization of basic public services must harness modern information technology. It can make public service organizations intelligent, improve their efficiency, and allow citizens to obtain basic public services in a more convenient way.

In addition, the modernization of basic public services should change the traditional bureaucratic culture and form a professional-based service culture. Only in this way can the characteristics of public service organizations be reflected.

CHAPTER 2

Macro Context for the Development of China's Public Services: Analysis and Prospects

2.1 Analysis of the Overall Vision and Goals of China's Modernization in 2030

This part shall begin with the official outlook and formalities. As early as 1987, Deng Xiaoping raised the famous Three-Step Development Strategy, with a clear aim "to quadruple per capita GNP to the level of moderately developed countries by the mid-21st century, bring the Chinese people an affluent life, and basically achieve modernization." The report of the 18th National Congress of the Communist Party of China (CPC) pledged "to attain the goal of completing the building of a moderately prosperous society in all respects by 2020." At the opening ceremony of the Boao Forum for Asia in April 2013, Comrade Xi Jinping, general secretary of the CPC Central Committee, called for "developing China into a great modern socialist country that is prosperous, strong, democratic, culturally advanced, and harmonious by the mid-21st century and realize the Chinese dream of the great rejuvenation of the Chinese nation." Moreover, according to the Communiqué of the Third Plenary Session of the 18th CPC Central Committee, "Decisive results must be achieved in key sectors, and a well-developed, scientific, procedure-based, and effective framework must be in place by 2020 to ensure institutions in all sectors be more mature." It can be inferred that in the Chinese government envisages China in 2030 as a modern and powerful country that is

K. Dong, N. Wei, *Transforming China's Public Services*,
https://doi.org/10.1007/978-981-99-3943-5_2

prosperous, democratic, socially harmonious, culturally advanced, and ecologically balanced. At that time, most of China's economic and social development indicators will reach the level of moderately developed countries.

Chinese researchers and research institutions in China have also made their analysis and predictions. In 2008, the China Center for Modernization Research of Chinese Academy of Sciences (CAS) released the *China's Modernization 2008: Research on International Modernization*, predicting that "in 2030, China's Objective Power Index will surpass Japan and rank second in the world."[1] In the *China's Modernization 2010: Overview of World Modernization* released in 2010, the research center made another prediction that China may reach the level of moderately developed countries around 2040 and achieve the Three-Step Development Strategy for China's modernization about 10 years ahead of schedule.[2] Justin Yifu Lin, a professor of economics at Peking University, estimated, "By 2030, China will be the most economically powerful country in the world."[3] According to the *Annual Report on Fiscal Policy of China 2010/2011*, released by the National Academy of Economic Strategy, CASS, in 2010, China will become the country with the largest aging population in the world by 2030.[4] Some researchers foresaw an urbanization rate of 80% in China by 2030.[5]

This topic has also been under the observation of international researchers and organizations. In 2011, Boston Consulting Group released a research report titled *A US$10 Trillion Challenge: A Playbook for China's Companies on Their Way to World-class*, predicting that, by 2030, China's GDP will reach US$17.7 trillion, with an increase of more than US$10 trillion. This growth is not driven by the large scale of cheap labor, but by increased investment and productivity.[6] In November 2013, Standard Chartered Bank released a special report on global economic growth prospects, arguing that the Chinese stock market will account for 20% of the

[1] China Center for Modernization Research, CAS, "China's national strength will surpass Japan and become the second largest in the world by 2030," *Northeast Electric Power Technology*, No. 3 (2008): 8.

[2] http://news.xinhuanet.com/fortune/201001/30/content_12903596.htm.

[3] Lin Yifu, "China's GDP to rank first in the world by 2030," *Brand*, No. 8 (2006): 83.

[4] http://www.chinanews.com/gn/2010/0910/2526415.shtml.

[5] Wan Guanghua, "2030: China's urbanization rate to reach 80%," *International Economic Review*, No. 6 (2011): 99–111.

[6] http://finance.ifeng.com/news/20110508/3994949.html.

value of the global stock market in 2030, second only to the United States in the world.[7] The most representative work is the research report *China 2030: Building a Modern, Harmonious, and Creative High-Income Society*, jointly compiled by the World Bank and the Development Research Center of the State Council of China. It predicted that even if its economic growth slows down, China will likely become a high-income country and the world's largest economy by 2030, successfully transitioning from a middle-income country to a high-income country.[8]

From the above analysis of China in 2020 and 2030 by institutions and researchers, some common points can be drawn. In 2020, China will remain the world's second-largest economy, with its per capita GDP reaching US$10,000, basically the level of moderately affluent countries and close to moderately developed countries. In 2030, China is very likely to surpass the US in terms of economic scale and become the world's largest economy, with its per capita GDP and national income into the ranks of middle- and high-income countries. However, in terms of human development index (HDI), national living standard, governance, social development, urbanization quality, social security, and ecological environment protection, China will still significantly lag behind developed countries, roughly close to or reaching the level of moderately developed countries or newly industrialized countries. In China's fundamental national philosophy and visions, China will be built into a prosperous, democratic, and culturally advanced modern socialist country. It will achieve scientific and sustainable development, pursue social fairness and justice, build a harmonious society, and reshape the international political and economic orders. The basic principles that China should uphold in public service provision in 2030 are full coverage, high quality, greater equality, moderate growth, and convenience and accessibility. On the one hand, it is necessary to steadily expand the coverage of public services and improve their quality. On the other hand, China must align public service reform with economic and social development and avoid the welfare trap caused by too-rapid development.

To sum up, the goal of China's development in 2030 is, in simple words, to approach or reach the level of moderately developed countries

[7] http://finance.sina.com/cn/stock/marketresearch/20131120/210817383138.shtml.

[8] World Bank and Development Research Center of the State Council, the People's Republic of China, *China 2030: Building a Modern, Harmonious, and Creative High-Income Society* (Beijing: China Financial and Economic Publishing House, 2013), 65.

or newly industrialized countries in economy, politics, society, science and technology, and the environment. This goal is the fundamental premise for target setting and path selection for the subtopics of education, medical care, aged care, culture, social services, and service purchase in the following parts of the research.

2.2 China's Macroeconomic Development in 2030

It is widely recognized that China has achieved the biggest and fastest economic growth in human history. From 1978 to 2014, the average annual growth rate of China's GDP was as high as 9.96%. China's GDP in 2014 was 174 times that of 1978.[9] In recent years, China has started to transform its economic development pattern and emphasize more on the quality of growth. In spite of this, after calculating China's GDP with the exchange rate approach, the World Bank's purchasing power parity (PPP) approach, or Angus Maddison's PPP (constant prices) approach, international organizations and Chinese researchers have reached an agreement that as long as China's economy can maintain a medium-high growth rate of 7% to 8% from 2011 to 2030, China's total GDP will reach around US$19.2 trillion in 2030. At that time, China's per capita GDP will exceed US$18,000,[10] basically reaching the level of moderately developed countries.

In terms of the development of economic and industrial structures, China's economic basis will turn from industry to the service sector, especially the modern service sector. In 2011, some researchers predicted that from 2015 to 2016, China's tertiary industry will overtake the primary and secondary industries to become the dominant industry. By 2020, the proportion of the tertiary industry will exceed 50%. By 2030, the tertiary industry will be an absolute dominator in China, accounting for 62.37%, with the proportion of primary and secondary industries at around 5.05% and 32.58%. It will also be the sector that hosts the largest number of jobs. In all, 51.8% of China's labor force will serve in the tertiary industry, and those in the primary and secondary industries will drop to 31.6% and

[9] National Bureau of Statistics of the People's Republic of China, *China Statistical Abstract 2015* (Beijing: China Statistics Press, 2015), 21.

[10] Li Shantong, Liu Yunzhong, et al., *China's Economy in 2030* (Beijing: Economic Science Press, 2011), 37. Hu Angang, Yan Yilong, and Wei Xing, *China in 2030: Towards Common Prosperity* (Beijing: Renmin University of China Press, 2011), 62.

16.6%.[11] In 2030, as China's tertiary industry becomes the leading industry, public services and industries, including education, medical care, aged care, and social services, will see rapid development and improvement.

In terms of national income and distribution, the per capita disposable income of urban households in China was CN¥343.3 in 1978, and that of rural households was CN¥133.6. These two figures increased to CN¥29,381.0 and CN¥9892.0, respectively, in 2014.[12] With the goal of "income doubling by 2020" raised by the 18th National Congress of the CPC, it is estimated that by 2020 and 2030, the per capita disposable income of urban households will reach CN¥60,000 and nearly CN¥100,000, respectively, and that of rural households will reach CN¥25,000 and nearly CN¥60,000, respectively. Moreover, the Engel coefficient of Chinese urban households decreased from 57.5% in 1978 to 35.0% in 2013, and that of rural households decreased from 67.7% to 37.7%.[13] At the current rate, it is expected that the Engel coefficients of urban and rural households will continue to drop to around 25% and 30%, respectively, in 2030. The decline of the Engel coefficient of urban and rural households indicates that Chinese urban and rural residents will lead a better life in the future. Urban and rural residents with their material life needs met will have greater demands for public services, such as education, medical care, and aged care, and will have more income for public service consumption.

In terms of employment, besides the potential significant changes in labor force distribution among the three industries mentioned above, changes will occur in the distribution of employed population between urban and rural areas. In 2014, China's urban employed population was 393 million, accounting for 50.9% of the total employed population, and the rural employed population was 379 million, accounting for 49.1%.[14] For the first time, the employed people in urban areas outnumbered those in rural areas. Considering the accelerated urbanization in China in the future, the proportion of urban employment must increase. It is estimated

[11] Hu Angang, Yan Yilong, and Wei Xing, *China in 2030: Towards Common Prosperity* (Beijing: Renmin University of China Press, 2011), 62.

[12] National Bureau of Statistics of the People's Republic of China, *China Statistical Abstract 2015* (Beijing: China Statistics Press, 2015), 62.

[13] National Bureau of Statistics of the People's Republic of China, *China Statistical Yearbook 2014* (Beijing: China Statistics Press, 2014), 158.

[14] National Bureau of Statistics of the People's Republic of China, *China Statistical Yearbook 2015* (Beijing: China Statistics Press, 2015), 112.

that by 2030, China's urban employed population, at more than 65% of the total population, will overwhelm that of rural agricultural employed population. Among an employed population of 850 million, the urban employed population will reach 550 million, while the rural employed population will be 300 million, dropping to 35%. In addition, in 2014, there were 9.52 million people in China's urban areas registered as unemployed, at a rate of 4.09%.[15] It is estimated that by 2030, the number of registered unemployed people in urban areas in China will be about 11 million, and the registered unemployment rate will remain at 3.6% to 3.9%. The increase of the employed population and changes in its structure will provide sufficient labor force for the public service industry. In addition, the distribution of the employed population will affect the focus and areas of future public service provision.

2.3 China's Urbanization and Population in 2030

New urbanization will be the only development path and a major development strategy for China in the upcoming 20 years. In 1978, China's urban population accounted for only 17.92% of the total.[16] Urbanization has been accelerated since the beginning of the Reform and Opening Up. In 2014, China's urban population (household-registered population and permanent residents) reached 749 million, accounting for 54.77% of the country's total, with an average annual growth rate of about 1.01%.[17] Future favorable policies, such as the government's promotion of new urbanization and rural land transfer, will further accelerate China's urbanization. With an average annual growth rate of 1.3% to 1.5%, it is expected that by 2020, China's urban population will occupy more than 60% of the total. By 2030, this proportion will increase to 68.9% to 71.7%, or even approach or exceed 72%, basically reaching the level of moderately developed countries. These basic numbers on urbanization indicate that the focus of China's basic public service provision will shift from rural to urban areas in the future. In terms of content and time of public service

[15] National Bureau of Statistics of the People's Republic of China, *China Statistical Yearbook 2015* (Beijing: China Statistics Press, 2015), 112.

[16] National Bureau of Statistics of the People's Republic of China, *China Statistical Yearbook 1996* (Beijing: China Statistics Press, 1996), 69.

[17] National Bureau of Statistics of the People's Republic of China, *China Statistical Yearbook 2015* (Beijing: China Statistics Press, 2015), 33.

provision, more consideration should be given to the characteristics of urban residents' work and life.

Urbanization is more than the increase of urban residents and the expansion of city size. Urban public facilities will develop and improve and people can enjoy urban life. In 2014, 97.6% of Chinese people had access to urban tap water, and 94.6% to gas. Every 10,000 people shared 12.99 public transport vehicles. The per capita urban road area was 15.34 m^2, and the per capita park green space area was 13.08 m^2.[18] Urban public facilities have been better off in general, but there are still problems and shortcomings. In particular, the number of public toilets per 10,000 people dropped to 2.79 in 2014 from 5.8 in 2000. At the current growth rate, which may be lifted as urbanization accelerates, public facilities in Chinese cities will be further improved. It is estimated that by 2030, almost everyone in China will have access to urban tap water and gas. Every 10,000 people will share 15.23 public transport vehicles (calculated by an average annual increase of 0.18 vehicle). The per capita urban road area will be 25.7 m^2 (calculated by an average annual increase of 0.63 m^2), and the per capita park green space area will be 16.55 m^2 (calculated by an average annual increase of 0.25 m^2).

China is a populous country, and its population will continue to grow, at least for a period, in the future. Since 2005, the natural growth rate of China's population has maintained an average of 4.8‰ to 5.5‰. At the end of 2014, China's total population was 1.368 billion; 51.24% of them were male, that is 701 million; and 48.76% were women, that is 667 million. As the two-child policy takes full effect, China's population growth will be stable and slightly upward, and the gender ratio will be evener. Accordingly, the UN predicts that China's population will exceed 1.4 billion by 2020 and reach its peak in 2030, with a total population of about 1.5 billion. After that, it will decline year by year, and the ratio of males and females will be roughly balanced, with about 755 million males and 745 million females.

Population aging will be a major challenge to China's population structure in the future. In 2014, the total number of people aged 65 and over reached 138 million, accounting for 10.1% of the total population.[19] The

[18] National Bureau of Statistics of the People's Republic of China, *China Statistical Yearbook 2015* (Beijing: China Statistics Press, 2015), 856.

[19] National Bureau of Statistics of the People's Republic of China, *China Statistical Yearbook 2015* (Beijing: China Statistics Press, 2015), 35.

Annual Report on Fiscal Policy of China 2010/2011, released by the National Academy of Economic Strategy, CASS, in 2010, pointed out that in the 30 years after 2011, China's population aging will see faster growth. By 2020, China's population aged 65 and above will increase to 13% of the total population. By 2030, this proportion will surpass that of Japan, and the proportion of the population aged 60 and over will increase by 16.55% annually. In 2040, the proportion of the population aged 60 and above will reach about 28%, making China the country with the largest aging population in the world. By 2050, the proportion of the population aged 60 and above will exceed 30%, and Chinese society will enter a stage of deep aging.[20] This situation means that China's demand for public services related to the elderly, such as medical care and aged care, will increase, and peak in around 2050.

2.4 China's Governance and Fiscal Situation in 2030

The Worldwide Governance Indicators (WGI), first released by the World Bank in 1996, is a project that assesses governance capacity of the governments of more than 200 countries around the world from six dimensions every year, namely voice and accountability, political stability and absence of violence/terrorism, government effectiveness, regulatory quality, rule of law, and control of corruption. Its underlying data sources include the assessment data published by 31 international organizations, as well as some World Bank survey data on household businesses, business information companies, and NGOs. These data are weighted and averaged with an unobservable component model to produce the final score. The standard normal unit for each indicator runs from -2.5 to 2.5 (with higher values corresponding to better governance), and the rank term ranges from 0 (the worst governance) to 100 (the best governance). We have worked on the data of 1996, 2002, 2007, and 2012, and shown China's WGI scores in the four years in Tables 2.1 and 2.2.

China's WGI results show that in the past two decades, the country scored high in government effectiveness, at the middle rank in regulatory quality, rule of law, and control of corruption, while lower in voice and accountability and political stability and absence of violence/terrorism. Compared with both itself and other countries, China has made progress

[20] http://www.chinanews.com/gn/2010/0910/2526415.shtml.

Table 2.1 Standard normal unit of China's WGI in 1996, 2002, 2007, and 2012

Dimension	*1996*	*2002*	*2007*	*2012*
Voice and accountability	-1.29	-1.57	-1.66	-1.59
Political stability and absence of violence/terrorism	-0.17	-0.36	-0.49	-0.55
Government effectiveness	-0.25	-0.05	0.19	0.01
Regulatory quality	-0.14	-0.53	-0.15	-0.25
Rule of law	-0.43	-0.41	-0.45	-0.49
Control of corruption	-0.25	-0.65	-0.59	-0.48

Table 2.2 Rank term of China's WGI in 1996, 2002, 2007, and 2012

Dimension	*1996*	*2002*	*2007*	*2012*
Voice and accountability	12.02	6.25	5.29	4.74
Political stability and absence of violence/terrorism	41.35	32.21	28.37	28.44
Government effectiveness	46.83	55.12	59.71	55.98
Regulatory quality	47.55	33.33	50.97	43.54
Rule of law	36.36	39.23	41.15	38.86
Control of corruption	44.90	33.66	33.50	38.76

and improvements in government effectiveness, regulatory quality, rule of law, and control of corruption. However, there seems to be a downward trend in voice and accountability, and political stability and absence of violence/terrorism.

Compared with other economic development data, WGI is inevitably Western-centrist and consciously or unconsciously applies the governance standards of some Western countries in evaluating the governance of developing countries. Therefore, it should be viewed a critically. However, WGI is still scientific and authoritative and can still provide some reference for China to improve its governance capacity.

By 2030, China's governance system and capacity will be basically modernized. The most significant frame of reference and standard for this modernization is to reach the governance level of moderately developed countries. Further analysis of WGI reveals the characteristics of WGI scores of some typical moderately developed countries (such as Greece, Portugal, South Korea, and South Africa), which are shown in Table 2.3.

The analysis of the above moderately developed countries' WGI scores shows that China still lags behind in the six dimensions of WGI. The gap in government effectiveness and regulatory quality is small, while that in

Table 2.3 Characteristics of the WGI scores of some moderately developed countries

Dimension	*Standard normal unit*	*Rank term*
Voice and accountability	≧0.55	≧60
Political stability and absence of violence/terrorism	≧0.05	≧45
Government effectiveness	≧0.3	≧60
Regulatory quality	≧0.5	≧58
Rule of law	≧0.3	≧65
Control of corruption	≧0	≧50

the remaining four dimensions is larger, especially in voice and accountability and rule of law. It can be seen that, with the steady improvement of China's governance in the future, especially the significant improvement of voice and rule of law, more favorable conditions will be created to reflect citizens' needs for public services and for the government to provide public services efficiently.

Since the beginning of the Reform and Opening Up, the fiscal revenue and expenditure of Chinese governments at all levels have grown rapidly. China's fiscal revenue increased from CN¥113.226 billion in 1978 to CN¥14,037.003 billion in 2014, with an average annual growth of 15.4%. China's fiscal expenditure increased from CN¥112.209 billion to CN¥15.178556 trillion, an average annual increase of 15.7%.[21] Notably, the state carried out the tax-sharing system reform in 1994, which focused on increasing fiscal revenue and adjusting the fiscal relationship between the central government and provinces, with the implementation of value-added tax as the main measure. Thanks to the reform, the proportion of the central government's fiscal revenue to the overall had increased from 20% to 30% before the reform to about 50%, reaching 51.1% in 2010.[22]

China's economic growth may slow down in the future, so will the growth of government revenues (calculated at an average annual growth rate of 10%), while that of government expenditures will remain unchanged (calculated at an average annual growth rate of 11%). In this case, the proportion of the central fiscal revenues and expenditures to the whole will increase (based on the central fiscal revenues accounting for 55% and

[21] National Bureau of Statistics of the People's Republic of China, *China Statistical Yearbook 2015* (Beijing: China Statistics Press, 2015), 214.

[22] National Bureau of Statistics of the People's Republic of China, *China Statistical Yearbook 2015* (Beijing: China Statistics Press, 2015), 214.

the central expenditures accounting for 25%). Based on the existing growth rate and the above factors, it is estimated that the fiscal revenues of Chinese governments at all levels in 2020 will be about CN¥24,493.06 billion, with the central government accounting for CN¥13,471.15 billion and local governments for CN¥11,021.91 billion. The fiscal expenditures will be about CN¥25,760.05 billion, with the central government accounting for CN¥6440.01 billion and local governments for CN¥19,320.04 billion. By 2030, the fiscal revenues of Chinese governments at all levels will be about CN¥63,528.17 billion, with the central government accounting for CN¥34,940.49 billion and local governments for CN¥28,587.68 billion. The fiscal expenditures will be about CN¥75,692.86 billion, with the central government accounting for 18,923.22 billion and local governments for CN¥56,769.64 billion.

The tax-sharing reform has significantly boosted the central government's revenue, allowing it to make more transfer payments to local governments, which has laid a solid foundation for China's fiscal system. However, it has brought various problems and challenges, including excessive spending, heavy debt burdens, and prominent land finance problems of local governments. For example, in 2014, the central government's fiscal revenues accounted for 45.9%, while its fiscal expenditures accounted for only 14.9%.[23] In the new era, China's fiscal system still needs further deepened reforms.

In the view of the research team jointly sponsored by the World Bank and the Development Research Center of the State Council, China's challenges in reforming its fiscal system over the next two decades will come in the four facets: (1) to contain the percentage of government expenditures in GDP and change their composition in response to China's new challenges, with higher allocations for social and environmental public investment and recurrent expenditures, (2) to improve the efficiency of revenue mobilization by changing in the structure of revenues and other means, (3) to reform intergovernmental fiscal relations by better aligning resource availability with the expenditure responsibility of different levels of government, and (4) to strengthen the management of government finances and improve the efficiency of public expenditures.[24]

[23] National Bureau of Statistics of the People's Republic of China, *China Statistical Yearbook 2015* (Beijing: China Statistics Press, 2015), 214.

[24] World Bank and Development Research Center of the State Council, the People's Republic of China, *China 2030: Building a Modern, Harmonious, and Creative High-Income Society* (Beijing: China Financial and Economic Publishing House, 2013), 65.

Taking into account the above factors of fiscal revenues and expenditures and structural changes, it can be expected that in 2030, the Chinese government will spend and invest more financial resources in basic public services. Expenditures on education will maintain a high proportion, while the proportion of investment in weak spots such as healthcare and general public services will increase substantially. It will provide considerable funds for the modernization of China's basic public services in 2030. In addition, the central government's functions in basic public service expenditures will be intensified, and support for local public services, especially in underdeveloped areas, will be strengthened with transfer payments and partner assistance.

2.5 Social Organizations in China in 2030

As China's urbanization and market-oriented reforms pick up the pace, people demand more professional and higher-quality social services in a larger scale. In addition, as the *danwei*/entity-based system has disintegrated, the primary social service provider has shifted from entities to the community organizations. Entities have declined, while community organizations are obviously insufficient, leaving space for social organizations to provide social services. Lessons from past public management reforms in developed countries show that social organizations are more flexible and efficient than government agencies in diversified, high-quality, and customized public service provision. Therefore, it is necessary to make judgments and predictions on the development of social organizations, as an important public service provider, in China in 2030.

The Chinese government divides Chinese third-sector organizations into social organizations, private non-enterprise entities, and foundations. The development of the three categories together epitomizes the development of third-sector organizations. Take social organizations as an example. In 1988, there were only 4446 registered social organizations in China. In 2014, the number reached 310,000, with an average annual growth rate of 251%.[25] Similarly, since the beginning of the Reform and Opening Up, especially 2000, the numbers of registered private non-enterprise organizations and foundations have risen sharply. In 1999, there were 5901 registered private non-enterprise organizations. By 2014, the number rose to 292,000, with an average annual growth rate of

[25] www.mca.gov.an/article/zwgk/mzyw/201506/20150200832371.shtml.

300.2%. The number of registered foundations in 2003 was 954, which increased to 4117 in 2014, with an average annual growth rate of 27.6%.

According to the statistics of the Ministry of Civil Affairs, in 2014, there were 606,000 registered social organizations of various types in China, and about 6.823 million people worked or served at social organizations. Statistics show that, if the 310,000 social organizations are categorized by areas, the largest number will be agriculture and rural development (60,202), followed by social services (44,630), industrial and commercial services (34,099), and culture (30,101). Those in small numbers are ecology and environment (6964), religion (4898), law (3270), and international and other foreign-related organizations (516).[26]

In 2011, there were 292,000 private non-enterprise organizations in China. Most of them are engaged in education (163,681), social services (42,244), health (23,404), and culture (14,148). Only a small number of them are engaged in ecology and environment (398), religion (82), and international and other foreign-related organizations (4).[27] Last, in 2014, there were 4117 foundations in China, including 1470 public foundations (35.7%), 2610 nonpublic foundations (63.4%), and 28 offices of overseas foundations (0.68%).[28]

Based on the above analysis of the growth and distribution of Chinese social organizations before 2014, as well as the development trend of Chinese social organizations, we make the following predictions for the development and distribution of social organizations across the country from 2020 to 2030.

2.5.1 The Number of Social Organizations Will Grow in Doubles

According to the growth rate before 2010 and the needs and characteristics of social development, the Ministry of Civil Affairs of China estimates that by 2020, the total number of social organizations will exceed 800,000. In 2030, the total number of registered social organizations across China will reach 1.693 million.

[26] www.mca.gov.an/article/zwgk/mzyw/201506/20150200832371.shtml.

[27] www.mca.gov.an/article/zwgk/mzyw/201506/20150200832371.shtml.

[28] www.mca.gov.an/article/zwgk/mzyw/201506/20150200832371.shtml.

2.5.2 More People Will Work at Social Organizations

It is estimated that the number of employees at social organizations will exceed 10 million by 2020 and exceed 20 million by 2030. Among them, about 200,000 will be qualified assistant social workers or social workers. The proportion will increase from the current 2.32‰ to 10‰. The workforce and service capabilities of social organizations will be significantly enhanced.

2.5.3 Social Organizations Will Have Much Better Fundraising Capacity and More Revenues

At the current development rate, it is estimated that by 2020 and 2030, the three types of social organizations that practice the accounting system for enterprises, administrative undertakings, and private nonprofit entities will significantly enhance their fund-raising capabilities, and their revenues and balances will increase drastically. This will provide a better foundation for social organizations to provide public services.

2.6 Characteristics of Social Organizations

In 2030, the total number of registered social organizations in China will reach 1.042 million. It is expected that the proportion of national and provincial organizations will further decline, while that of prefecture- and county-level organizations will increase to about 95%. The number of industry organizations will reach 210,000, exceeding 20%. Influenced by the future development of China's legal system construction, democracy, urbanization, and scientific and technological research, it is expected that the registered service organizations will grow steadily, with legal and civil rights-defending organizations increasing rapidly. In terms of proportion, the proportion of social service organizations and technology and research organizations will increase significantly. Public service social organizations for education, healthcare, ecology and environment, culture, and sports will grow steadily. The number of international, foreign-related, legal, and religious organizations will increase significantly but still occupying a small proportion. The proportion of agriculture and rural development and industrial and commercial service organizations will decrease.

2.7 Characteristics of Private Non-enterprise Entities

In 2030, the total number of registered private non-enterprise organizations in China will reach 648,000. By then, the number of individual and cooperative organizations registered as private will exceed 320,000, with the proportion rising significantly to nearly 50%. The proportion of legal-person organizations will remain around 50%. Influenced by the future development of China's legal system construction, democracy, urbanization, and scientific and technological research, private non-enterprises organizations for education, social services, and healthcare will still dominate. Ecology and environment, sports, science and research, law, and international and foreign-related organizations will maintain their upward momentum. The proportion of agriculture and rural development and industrial and commercial services will show a downward trend.

2.8 Characteristics of Foundations

In 2030, the total number of registered foundations in China will reach 57,000. Among them, the proportion of public foundations is estimated to rise to about 60%, that of representative offices of overseas foundations will rise to 10%, and that of nonpublic foundations will drop to about 30%. In terms of area, organizations for social services will dominate at 35%, closely followed by organizations for education, culture, and healthcare. The proportion of foundations for sports, law, and international and foreign-related affairs will increase slightly. The proportion of agriculture and rural development, industrial and commercial services, occupational and professional organizations will decline.

CHAPTER 3

Modernization of the Basic Public Education Service System

3.1 Modernization of the Basic Public Education Service System: An Interpretation Based on Policies

3.1.1 Background of the Construction of the Basic Public Education Service System

Basic public education services refer to the basic public services provided in education, which are public funded and provided mainly by the government in diversified forms. They are public, inclusive, fundamental, and aimed at development. The public education services that are most concerned by and most directly related to the people, and most practical and most vital to the people, are the basic premise and foundation for people's lifelong development, and the fundamental driving force and core of social stability and sustainability.

Since the beginning of Reform and Opening Up, China has seen steady economic growth, better lives of the people, more extensive and deeper internationalization, and advancements in government institutional reforms. Against such a backdrop, the Chinese government has been aware of that education is to the improvement of comprehensive national strength and crucial to national security. *Learning to Be*, a report published by the United Nations Educational, Scientific and Cultural Organization (UNESCO) in 1972, pointed out, "Now, probably for the

K. Dong, N. Wei, *Transforming China's Public Services*,
https://doi.org/10.1007/978-981-99-3943-5_3

first time in the history of humanity, development of education on a world-wide scale is tending to precede economic development." It has revealed the latest trend in education worldwide. Education development has been prioritized in many countries' national strategy, showing an irresistible momentum. China has begun to draw advanced educational experience and theories in and outside of the country and prioritized education development practically. The increasing significance of education development is also in line with the development of institutional reform and government transformation. Moreover, people's education is closely related to the relocation of industrial centers; the degree and quality of people's education set the direction for the country's future development.

The Outline of the National Plan for Medium and Long-Term Education Reform and Development (2010–2020) clearly states that in 2020, China should develop fair education that benefits everyone, build a basic public education service system covering urban and rural areas, be on the way to equalized basic public education services, and narrow regional disparities. Education priorities will shift from educating the elite to educating the masses. Education will be an important area, crucial to constructing the national public service system, and essential to the public service system. The Plan for the Development of the National Basic Public Service System during the 12th Five-Year Plan Period explicitly prioritizes the construction of the basic public education service system and puts education on the top over many other basic public services. The Plan defines the scope of basic public education services during the 12th Five-Year Plan period, including inclusive preschool education, nine-year compulsory education, and high school education into basic public education services. It also defines the standards for basic public education services during the 12th Five-Year Plan period, which is an important reference for local governments at all levels to effectively provide basic public education services.

Social development supports the national strategy of prioritizing education development. The effect of education on economic growth is slow but huge. On the other hand, education development cannot exceed the development stage of society. Since the beginning of Reform and Opening Up, as China has shifted from seeking survival to seeking development in an all-round way, the demands of economic and social development on education and the types and calibers of human resources have changed profoundly. People have more urgent demands for high-quality basic public education services and more demands for personalized and featured education closely related to their personal development. As urbanization picks up pace, there is an urgent need to improve the basic public education service system covering both urban and rural areas, a more adaptive education model. The construction of the education public service system is closely related to the future development direction of China's economy and popular opinion, as well as the upgrading of the socialist market economic system and sustainable, stable, and healthy social progress.

Since the beginning of Reform and Opening Up, China's education reform has made remarkable achievements. However, there is still contradiction between social development and education reform. The demand for basic public education services is growing in an all-round way and drastically changing structure, while basic public education services are not in place. It has become a prominent contradiction in the construction of the basic public education service system, affecting and restricting the transformation of China's economic development patterns.

Under the guiding policy of prioritizing education development, the modernization of China's basic public education service system is mainly based on two ideas. The first is to return to the essence of education and take education as the main approach to cultivate all-round talents. The second is to integrate national strategic needs into the public service system based on education's effect on the country, to organically combine personal worth and national strategic goals. The key to a basic public education service system lies in stability and balance. We will explore the "China model" of modernizing China's basic public education service system in three aspects: concept construction, financial support, and the evaluation mechanism.

3.1.2 Connotation of the Modernization of Basic Public Education Services

At the current progress and speed of economic growth, China will basically achieve the goal of "all our people enjoying their rights to education" in 2020. On the basis of consolidating compulsory education, China will equalize compulsory education and achieve basic balance among provinces. China's basic public education service system is shifting from pursuing quantity to pursuing quality, or from education availability to quality education. China is taking various measures to ensure the quality and balance of formal school education.

In the 11th Five-Year Plan period, there are two major tasks in education. The first is to improve the appeal of secondary vocational education. By furthering industry-education combination and school-enterprise cooperation, a vocational school-running mechanism led by the government, oriented by the industry, and involving enterprises will be established to strengthen the basic capacity of secondary vocational education. Free secondary vocational education for students from urban and rural families with financial difficulties and agriculture students will be adopted, making secondary vocational education more appealing to them. The second is to address the difficulties in kindergarten admission in preschool education, including the problem of preschool children attending school in urban areas and preschool education in rural areas.

In 2020, China will basically solve the aforementioned problems. As education becomes more popular and solid, we need to shift the focus of education reform and development from quantity to quality. How to effectively improve quality will certainly be a crucial part in the "China model" exploration. The basic orientation is to "meet the diverse needs of high-quality education through quality improvement on the basis of strong and balanced education."

First, we should ensure the enrollment rate of compulsory education while improving its quality. We should adjust the instruction content and methods of compulsory education, focusing on cultivating students' basic abilities. Moreover, we should increase public funding for high school and preschool education and, where possible, make them compulsory. On the other hand, we should better support education reform by making it into law, so that education can better adapt to socioeconomic development and individuals' all-round development and help citizens realize self-worth, as individual development will promote of the society's overall welfare.

Second, after ensuring balance, we should determine the priority in education development. We should consider China's national conditions and traditions, highlight Chinese characteristics, and extend the priority of education development beyond both ends of contemporary compulsory education, that is, pre-schooling and high-schooling. We should improve education quality to meet education needs. Specifically, we should allocate more high-caliber teachers in preschool education to meet basic education needs in terms of scale and quality. We should also consider to integrate special education into general education, adjust the development philosophy for special education, and improve the education philosophy for all.

Third, we should implement streaming in junior high school. The "higher vocational school-high school" dual-track education model adapts to China's large-scale industrialization, offers the "livelihood education" needed by the people, and provides individuals with skills to earn a living. We should cultivate professional skills and professionalism among the people and further popularize industrial civilization. Around 2020, there will be fundamental changes in the interaction between China, as the world's largest economy, and other economies. Therefore, it is necessary to make visionary adjustments to the education structure according to the industrial structure.

Fourth, we should extend basic public education services according to national strategies and include higher vocational education into the public service system. The central government has put forward an innovative concept, "education leading public services," and planned for the rapidly increasing demand for talents in many public service sectors. As the public service system further grows, there will be more demands for professionals and skill training. Such demands will come from traditional public service areas, including social security, medical care, family planning, housing security, culture and sports, and other areas closely related to people's living environment, including transportation, communications, public amenities, environmental protection, public safety, consumer security, and defense security.

Fifth, we should redefine government functions and reform the responsibility assignment of the central and local governments in basic public services. Specifically, we should manage and coordinate government functions in a pyramid structure, and promote the administration-to-service transformation. We should fully exert the central government's macro-management functions and local governments' initiative, carry out regional-specific public service provision reforms, and integrate education services with regional collaborative innovation to facilitate entrepreneurship and innovation. We should also restructure the public funding system. In the areas where education services can be purchased, we should use diversified and flexible methods (education vouchers, entrusted agents, government contracts, etc.) to allocate public resources, and seek the optimal model for educational resource allocation.

Sixth, in the provision of basic public education services, education services provided by the government must clearly define the "civil servant" status of their subjects, including teachers, principals, and administrators. They are subject to centralized deployment, with clearly defined rights, responsibilities, and interests. Moreover, we should incorporate basic public services into the purview of government administration, provide necessary administration authority, and implement it through appropriate mandatory measures. The reform of teachers' status will be an effective breakthrough for basic public education services and the foundation for a stable and sound service system.

Seventh, we should link information technology closely with China's education development and use information technology to bolster the modernization of the basic public service system and the development of educational services. We should integrate instruction and education systems with information technology to create smart basic public education services.

3.2 Progress and Problems in the Construction of the Basic Public Education Service System

3.2.1 *Progress in the Construction of the Basic Public Education Service System*

Overview

In April 2014, the State Council released the mid-term evaluation report of the 12th Five-Year Plan, which recognized the overall education development in the period. Education, as a priority, maintained a good momentum and was more popularized. Public investment in education was further increased, and key education development projects advanced steadily. Progress was made in education reform and development, equal access to education was fruitful, public educational resources were better allocated, and vocational education and skills training were expanding. In 2012, China's GDP was CN¥51,894.211 billion, and the state fiscal education funding accounted 4.28%, an increase of 0.35 percentage points (pp) over the previous year, reaching the goal set in the 12th Five-Year Plan.

As shown in Table 3.1, China's educational expenditure has increased year by year. Its total investment in education in 2012 doubled from 2008. The share of public budgetary education spending in GDP has increased year by year, and exceeded 4% for the first time in 2012.

In 2012, the gross enrollment rate of China's three-year preschools was 64.5%, an increase of 7.9 pp over 2010. The completion rate of nine-year

Table 3.1 Fiscal education spending

Year	*Total education spending (CN¥100 million)*	*Growth rate of total education spending (%)*	*Fiscal education spending (CN¥100 million)*	*Growth rate of fiscal education spending (%)*	*Fiscal education spending to total (%)*	*Public budgetary education spending to fiscal expenditure (%)*	*Share of public budgetary education spending in GDP (%)*
2008	14,500.74	19.37	10,449.63	26.20	72.06	16.32	3.48
2009	16,502.71	13.81	12,231.09	17.05	74.12	15.69	3.59
2010	19,561.85	18.54	14,670.07	19.94	74.99	15.76	3.65
2011	23,869.29	22.02	18,586.70	26.70	77.87	16.31	3.93
2012	27,695.97	16.03	22,236.23	19.64	80.29	18.33	4.28

compulsory education was 91.8%, an increase of 2.1 pp over 2013. The gross enrollment rate of senior high schools was 85%, an increase of 2.5 pp over 2013; that of higher education was 30%, an increase of 3.5 pp. Focusing on high-caliber and highly skilled talents, China put more efforts in training personnel of different disciplines. In 2012, the total number of professionals and technicians in the public sector was 29.77 million.

Despite the remarkable achievements, the proportion of public funds per student in China is comparatively low (see Table 3.2). Public budgetary education spending in general primary school, general junior high school, and secondary vocational education is mostly for personnel, including the salary of teachers, school managers, and other school staff. The public part, including the investment in students, only accounts for about 30%, which needs to be increased.

Progress by Category

The Current Scope and Scale of China's Basic Public Education Services

According to the development plan for China's basic public service system during 12th Five-Year Plan period, the government's basic public education services include (1) free nine-year compulsory education for school-age children and adolescents, free accommodation for rural boarding

Table 3.2 Public budgetary education spending and public funding per student and their growth rate in 2012

Education Degree	*General primary school*	*General junior high school*	*General senior high school*	*Secondary vocational school*	*General higher education*
Education spending (CNY)	6128.99	8137.00	7775.94	7563.95	16,367.21
Y-o-y increase (%)	26.30	24.38	29.61	23.03	17.94
Public funding (CNY)	1829.14	2691.76	2593.15	2977.45	9040.02
Y-o-y increase (%)	33.86	31.63	53.66	34.55	21.19
Public funding/ education spending (%)	29.84	33.08	33.35	39.36	55.23

students in compulsory education, and living allowances for students from families with financial difficulties; (2) better nutrition for rural compulsory education students in poverty-stricken areas; (3) free secondary vocational education for rural students, urban students with financial difficulties, and agriculture students; (4) funding for students from families with financial difficulties to receive general high school education; and (5) funding for children from families with financial difficulties, orphans, and children with disabilities to receive preschool education. The coverage of basic public education services in China in the end of 2015 is as follows.

Preschool Education

There are 223,700 kindergartens nationwide, an increase of 13,800 over 2014. There are 42.6483 million children attending kindergartens (including affiliated programs), a y-o-y increase of 2.1411 million. The number of kindergarten directors and teachers totals 2.3031 million, a y-o-y increase of 222,800. The gross enrollment rate of preschool education is 75.0%, an increase of 4.5 pp over 2014.

Compulsory Education

The national completion rate of nine-year compulsory education is 93.0%. The number of full-time teachers is 9,160,800. There are 242,900 compulsory education schools nationwide, a y-o-y decrease of 11,100. Among them, there are 190,500 primary schools, a y-o-y decrease of 10,900; and 52,400 junior high schools, a y-o-y decrease of 218.

A total of 31,400,700 students are enrolled in compulsory education nationwide, among which 17,290,400 are enrolled in primary schools, and 14,110,200 in junior high schools. There are 140 million students in compulsory education, with 96,921,800 primary school students and 43,119,500 junior high school students. 2015, there are 14,372,500 primary school graduates and 14,175,900 junior high school graduates.

The net enrollment rate of primary school-age children is 99.88%. The gross enrollment rate of junior high school education is 104.0%, and the enrollment rate of junior high school graduates is 94.1%.

The construction area of general primary schools (including teaching venues) is 673,520,400 m^2, a y-o-y increase of 26,548,500 m^2. The proportions of schools with facilities and apparatus up to the standard are as follows: 64.5% with up-to-standard sports venues (centers), 68.9% with up-to-standard sports apparatus, 67.9% with up-to-standard musical instruments, 67.6% with up-to-standard fine arts paraphernalia, and 69.0% with up-to-standard math and natural sciences experiment apparatus.

The construction area of junior high schools (including teaching venues) is 550,420,700 m^2, a y-o-y increase of 24,785,200 m^2. The proportions of schools with facilities and apparatus up to the standard are as follows: 78.7% with up-to-standard sports venues (centers), 83.6% with up-to-standard sports apparatus, 82.3% with up-to-standard musical instruments, 82.0% with up-to-standard fine arts paraphernalia, and 85.9% with up-to-standard math and natural sciences experiment apparatus.

Special Education

There are 2053 special education schools nationwide, a y-o-y increase of 53, staffed by 50,300 full-time teachers. A total of 83,300 special education students are enrolled across the country, a y-o-y increase of 12,600, making the total number of students 442,200, a y-o-y increase of 47,400. Among them, 36,700 are with visual disabilities, 89,400 with hearing disabilities, 232,100 with intellectual disabilities, and 84,000 with other disabilities. 2015, the number of special education graduates was 52,900, a y-o-y increase of 3900.

There are 44,800 special education students enrolled in general primary and junior high schools or their special education programs, making the total number of students 239,600, accounting for 53.7% of the total number of students enrolled in special education and 54.2% of the total number of special education students.

Senior High School Education

There are 13,200 general senior high schools nationwide, with 7,966,100 students enrolled, a y-o-y increase of 106. The total number of students is 23,744,000, a y-o-y decrease of 260,700. This year, there are 7,976,500 graduates, a y-o-y decrease of 19,700.

The construction area of senior high schools is 471,359,600 m^2, a y-o-y increase of 17,899,400 m^2. The proportions of schools with facilities and apparatus up to the standard are as follows: 87.1% with up-to-standard sports venues (centers), 88.8% with up-to-standard sports apparatus, 97.5% with up-to-standard musical instruments, 87.6% with up-to-standard fine arts paraphernalia, and 89.8% with up-to-standard math and natural sciences experiment apparatus.

Secondary Vocational Education

There are 11,200 secondary vocational schools nationwide, a y-o-y decrease of 676. A total of 6,012,500 students are enrolled in secondary vocational education, accounting for 43.0% of the total enrollment in senior high school education, a y-o-y decrease of 185,100.

There are 16,567,000 students in secondary vocational education, accounting for 41.0% of the total number of students in senior high school education, a y-o-y decrease of 985,800. The number of graduates from secondary vocational education this year is 5,678,800, a y-o-y decrease of 550,600. Secondary vocational schools have a total of 1,101,800 teaching staff, a y-o-y decrease of 30,300. Among them there are 844,100 full-time teachers, a y-o-y decrease of 14,300.

Resource Allocation in Basic Public Education Services

Basic public education service resources in China are unevenly allocated in general. They are concentrated in the southeast coast and the western and northern border areas, while the central hinterland finds less distribution, which indicates disparities between regions. On the other hand, urban areas are richer in resources than rural areas, and the gap between urban and rural areas is still large.

According to the China Educational Finance Statistical Yearbook 2012, the education spending per student in Beijing, Shanghai, Jiangsu, Inner Mongolia, and Tibet was over CN¥10,000, while that in Henan, Hubei, Jiangxi, Guizhou, and Guangxi was less than CN¥4600. Regarding the whole country in 2012, the institutional education spending per primary school student in the central-western China was higher than that in the central region.

An analysis of education spending per student at all degrees in all Chinese provinces and cities in 2011 shows that the region with the highest education spending per student is Beijing, while the provinces with the lowest are Henan, Guangxi, Guizhou, Jiangxi, and Hubei. These provinces, except Henan, are all in the central-western China (see Table 3.3). The gap between the highest and the lowest is enormous, even taking into account the difference in consumption, which does no good to educational equality.

As shown in Table 3.4, education spending per student at all degrees in Chinese urban areas is higher than that in rural areas, while that in rural areas is lower than the average, although the gap is inconspicuous. However, in the comparison of the lowest values in Table 3.5, the gap becomes significant. Although some well-developed rural areas can reach the average, the gap between poverty-stricken rural areas and other areas is still enormous.

Table 3.3 Regional education spending per student at all degrees in 2011

Education degree	*National average (CNY)*	*Beijing (CNY)*	*Minimum (CNY)*	*Beijing / Minimum (times)*	*Beijing / National average (times)*	*Minimum / National average (times)*
Kindergarten	5374	18,031	1896 (Guangxi)	9.51	3.35	0.35
General primary school	7923	24,316	3294 (Henan)	7.38	3.06	0.41
General junior high school	10,135	34,602	4513 (Guizhou)	7.66	3.41	0.44
General senior high school	11,947	41,349	6401 (Henan)	6.45	3.46	0.53
Vocational senior high school	10,954	34,843	5716 (Jiangxi)	6.09	3.18	0.52
Special school	40,952	86,034	22,329 (Hubei)	3.85	2.10	0.54
Secondary vocational school	11,488	29,248	6636 (Hubei)	4.40	2.54	0.57

Table 3.4 Comparison of average education spending per student at all degrees in 2011: Rural areas vs. overall

Education Degree	*Overall average (CNY)*	*Average in rural areas (CNY)*	*Overall average / Average in rural areas*
Kindergarten	5374	2988	1.79
General primary school	7923	7902	1.00
General junior high school	10,135	10,024	1.01
General senior high school	11,947	9499	1.25
Vocational senior high school	10,954	9211	1.18

Standards and Approaches of Basic Public Education Services

China has few statutory standards for the provision of basic public education services, only the Management Standards for Compulsory Education Schools (for Trial Implementation) introduced in August 2014. There are

Table 3.5 Comparison of the lowest education spending per student at all degrees in 2011: Rural areas vs. overall

Education degree	*Overall minimum (CNY)*	*Minimum in rural areas (CNY)*	*Overall minimum / Minimum in rural areas*
Kindergarten	1896	575	3.29
General primary school	3294	3154	1.04
General junior high school	4513	4251	1.06
General senior high school	6401	4923	1.30
Vocational senior high school	5716	2290	2.49

Table 3.6 Pupil-to-teacher ratio at different education degrees in China (the number of teachers = 1)

Indicator	*2013*	*2012*	*2011*	*2010*	*2009*	*2008*
Primary schools	16.76	17.36	17.71	17.70	17.88	18.38
Junior high schools	12.76	13.59	14.38	14.98	15.47	16.07
General senior high schools	14.95	15.47	15.77	15.99	16.30	16.78
Vocational senior high schools	22.97	/	21.59	23.66	23.65	23.47
General secondary vocational schools	/	/	31.53	26.10	27.82	31.27
Colleges and universities	17.53	17.52	17.42	17.33	17.27	17.23
Undergraduate colleges and universities	/	17.98	17.48	17.38	17.23	17.21
Three-year colleges	/	17.33	17.28	17.21	17.35	17.27

few laws and regulations in this regard, and lack practical approaches, incentives, and constraints.

The pupil-to-teacher ratio is a key indicator to measure education development., Table 3.6 presents the ratio at all education degrees in China according to the Statistical Bulletin on National Education Development in 2014. The ratio sees slight changes, but in compulsory education and senior high school education, it decreases year by year. Among them, that in junior high school is the lowest.

The staffing of primary and high schools is still governed by the standards introduced ten years ago. The standards are not grounded on the reality that the number of teachers trained is increasing, while the number of new students is declining. It needs to be considered and redeveloped.

In 2007, the Ministry of Education issued the Opinions of the Ministry of Education on Strengthening the Teaching Staff at Secondary Vocational

Schools in the 11th Five-Year Plan Period, anticipating that in 2010, the number of teachers in secondary vocational schools nationwide would reach 1.3 million, of which part-time teachers would account for 30% of the total, and the pupil-to-teacher ratio would grow to about 16:1. By this standard, the pupil-to-teacher ratio is relatively high at this stage.

Regarding preschool education, the Ministry of Education issued the Interim Standards for the Staffing of Kindergartens in 2013. It stipulates that the staff-children ratio in new full-time kindergartens nationwide must be 1:7 to 1:5, and the existing full-time kindergartens should meet the requirements of this standard within three years.

Rural areas are the focus of the equalization of basic public services. Therefore, it is necessary to improve rural production and living conditions. We should establish a sound mechanism for urban-rural integrated development, accelerate the removal of institutional obstacles that restrict the coordinated development of urban and rural areas, and balance allocation of public resources and the free flow of production factors between urban and rural areas. We should coordinate urban and rural development planning; promote the integration of urban and rural infrastructure, public services, and social management; and improve the equitable exchange of production factors between urban and rural areas. We should allocate more state fiscal expenditure and budgetary investment in fixed assets to agriculture and rural areas. In addition, we should "expand the coverage of public finance in rural areas, comprehensively strengthen financial support for rural public services, improve the quality and balanced development of rural compulsory education, advance free rural secondary vocational education, and actively develop rural preschool education."

Provision of Basic Public Education Services

Basic public education services follow the principles of "led by the government, expanding coverage, focusing on equity, and improved efficiency." On the basis of guaranteeing compulsory education, it focuses on protecting the poor and vulnerable, committed to equalization and accessibility of basic public services. In terms of central and local investment, local governments were taking on heavier burdens in the past five years due to the administration system of China's education, especially basic education, which can be summarized as "local government taking responsibility and running schools and administering education at levels." As shown in Table 3.7, the proportion of local funding has remained above 80%. Since the introduction of the tax-sharing system in China in 1994, the scale of

Table 3.7 Comparison of central and local education spending

Year	*Fiscal education spending of the state (CN¥100 million)*	*Public budgetary education spending (CN¥100 million)*	*Central government's fiscal education spending*		*Local government's fiscal education spending*	
			Amount (CN¥100 million)	*Percentage (%)*	*Amount (CN¥100 million)*	*Percentage (%)*
2008	10,449.63	9685.56	1603.7	16.56	8081.85	83.44
2009	12,231.09	11,419.30	1981.39	17.35	9437.91	82.56
2010	14,670.07	13,489.56	2547.34	18.88	10,942.22	81.82
2011	18,586.70	16,804.56	3268.59	19.45	13,535.97	80.55
2012	22,236.23	20,314.17	3781.55	18.62	16,532.62	81.38

Table 3.8 Education spending per student at all degrees in 2011

Education Degree	*National average (CNY)*	*Rank*	*Beijing (CNY)*	*Rank*	*Minimum (CNY)*	*Rank*
Kindergarten	5374	7	18,031	7	1896 (Gguangxi)	7
General primary school	7923	6	24,316	6	3294 (Henan)	6
General junior high school	10,135	5	34,602	4	4513 (Guizhou)	5
General senior high school	11,947	2	41,349	2	6401 (Henan)	3
Vocational senior high school	10,954	3	34,843	3	5716 (Jiangxi)	4
Special school	40,952	1	86,034	1	22,329 (Hubei)	1
Secondary vocational school	11,488	4	29,248	5	6636 (Hubei)	2

the central finance has expanded, the "two proportions"[1] have grown, and funds available to local governments are limited. In short, local governments have greater administrative power, while the central government has greater financial power. Financial input for basic public education services mainly comes from local governments. Due to the significant gap in economic development between regions, the central government should use institutionalized transfer payments to promote the equalization of basic public education services.

Different degrees and types of education have their own characteristics and thus cost differently. As shown in Table 3.8, in generally, the higher the education degree, the higher the education expenditure. Among different types of education, special education is the costliest, significantly higher than that of other types of education. Vocational education costs less than general education, which is related to the fact that most Chinese people think more highly of general education.

Preschool education receives the lowest spending per student. Although China's preschool education has attracted attention and emphasis from parents, it has not yet been included in the public service system, which makes the spending for it arbitrary and not compulsory. Market regulation

[1] The "two proportions" refer to the proportion of fiscal education spending to GNP and the proportion of education spending to the local government's total fiscal expenditure.

of the content of basic public services is prone to many problems and requires further revision of regulations and continuous and sufficient government investment.

China's spending in special education far exceeds that in other categories, which shows the state's extraordinary care for vulnerable groups. In 2010, under the theme of "caring for and supporting special education," the central government improved the special education system, increased professional training for teachers, expanded the scale of "learning in regular classroom (LRC)" and special education programs in general schools, and supported renovating and building special education schools in the central and western China. The theme for 2011 was "guaranteeing the right to education for children and adolescents with disabilities." The year 2012 was a continuation of the theme of 2010 and the work in the previous two years, with efforts in incorporating some work objectives into policies. The theme for 2013 was "supporting special education." The central government launched some key projects and began the decisive initiative for compulsory education for children and adolescents with disabilities. The central government also launched key construction projects for special education to equip backward special schools with basic teaching and rehabilitation apparatus and support general schools to build classrooms for students with disabilities.

Education spending is mainly financed by the state's fiscal education input. China's total education spending is composed mostly of the state's fiscal education spending, private school sponsors' input, social donations, and institutional income. Figure 3.1 shows the composition of China's education spending in 2011. Among them, the state's financial education spending accounts for 77.87%, while the private investment in private education and social donations accounts for less than 1% in total. The Outline of China's Education Reform and Development issued in 1993 advocated "encouraging the private sector to run and donate to schools." Twenty years have passed, but the private sector's investment in education is still a fraction. The 12th Five-Year Plan mentions to further define the responsibilities of the central and local governments, and strengthen the provincial overall planning of education. The Plan encourages and guides the private sector to establish schools, guarantees the equal legal status of private and public schools, and regulates school operation.

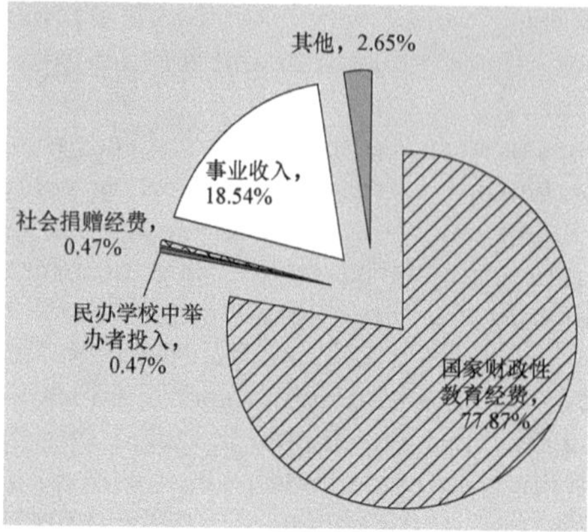

国家财政性教育经费	The state's fiscal education spending
民办学校中举办者投入	Private school owners' input
社会捐赠经费	Social donations
事业收入	Institutional income
其他	Other

Fig. 3.1 Composition of China's education spending in 2011

3.2.2 Problems and Challenges in the Construction of the Basic Public Education Service System

Limited Scope, Quality, and Standards of Basic Public Education Services

Scope to Be Expanded

From the end of World War II to the 1980s, the construction of basic public education service systems in Western developed countries showed a trend of "focusing on compulsory education and extending beyond both ends." Compulsory education has been extended, and preschool education and part of vocational education have been included in public education services.

In the twenty-first century, the Chinese government has put extra efforts in building the basic public education service system. The Plan for the Development of the National Basic Public Service System During the 12th Five-Year Plan Period issued in 2012 includes inclusive preschool education, nine-year compulsory education, and senior high school education into basic public education services, but without vocational education, education in ethnic-minority areas, and education for vulnerable groups.

Lessons from many countries show that if a country includes vocational education, especially secondary vocational education, into basic public education services, it will see economic growth and take the lead in the international division of labor. Vocational education is directly related to China's strategy to develop a quality workforce. In Australia's view, future prosperity is closely related to the skills and productivity of the workforce. Therefore, vocational education, especially secondary vocational education, is highly essential. In Germany's view, vocational education is the "secret weapon" for economic boom and coping with economic and social changes, and developing vocational education falls in the federal government's responsibility. The EU has developed the Europe 2020 strategy to vigorously promote the modernization and integration of vocational education. In China, while the nine-year compulsory education is basically universal, the weak development of vocational education, especially secondary vocational education, has become increasingly prominent. Solving the problem will directly boost the transformation of China's economic development pattern and industrial structure optimization.

Education in China's ethnic-minority areas is still a vulnerable spot in China's education development, restricted by scarce educational resources, especially teachers. The inclusion of ethnic-minority areas in basic public education services with priority support is related to the country's overall stability. It is also an important part of implementing the CPC and the central government's socialist ethnic-minority policies.

In the inclusion of the education for vulnerable groups in public education services, it is necessary to ensure the right to education of vulnerable groups such as people with disabilities and in remote and impoverished areas, migrant workers' children living with their parents, and rural left-behind children. It is important to fully understand that the purpose of basic public education services is not only to build comprehensive public services but also to compensate vulnerable groups to better ensure education equity, especially equal opportunity.

Table 3.9 Share of fiscal education spending in GDP in OECD countries (2009) and China (2011)

Country	*Preschool education spending / GDP (%)*	*Primary, secondary, and postsecondary education spending / GDP (%)*	*Higher education spending / GDP (%)*	*National education spending / GDP (%)*
Australia	0.1	3.8	1.1	5.0
Canada	/	3.2	1.8	5.1
France	0.7	3.9	1.3	5.9
Germany	0.4	3.2	1.3	5.1
Japan	0.1	2.7	0.8	3.8
U.K.	0.3	4.5	0.8	5.6
U.S.	0.3	3.9	1.3	5.5
Average of OECD countries	0.6	3.8	1.4	5.8
China	0.1	2.4	0.9	3.9

Source: Wang Jian, "Governance allocation and expenditure responsibility division of basic public service: A case of OECD countries," *Journal of University of Electronic Science and Technology of China (Social Sciences Edition)*, 04 (2013): 49–55

To conclude, it is imperative to include vocational education (including secondary vocational education, higher vocational education for agriculture, rural areas, and farmers and industries in short supply), education for vulnerable groups, education in ethnic-minority areas, preschool education, and senior high school education into China's basic public education service system. It is the premise and key to promoting China's basic public education service system to define the scope of basic public education services and improve it through legislation.

Quality to Be Enhanced

In the twenty-first century, most governments around the world have increased their support for the construction of public education service systems. In 2000, the world's average share of public education spending in GDP was 4.2%, which went up to 4.5%[2] in 2007. The figure in OECD countries was 5.8%, while in China, only 3.9% (see Table 3.9). China's basic public education services are relatively low in quality and investment.

[2] Guo Qingwang, Liu Xiaolu, Jia Junxue, "State financial capabilities and basic public education service system construction in the national 12th Five-Year Plan," *Renmin University of China Education Journal*, 01 (2011): 35–47.

It is necessary for the state to increase fiscal support for constructing the basic public education service system. On the other hand, it is necessary to establish a measurement index for basic public education services for tracking and assessing China's basic public education services and prompt disclosure to the public.

Standards to Be Improved

Standards for the basic public education service system refer to technical and management specifications for basic public service activities designed to achieve the goals in a certain period. Many countries are developing quality assessment standards for basic public education services. The U.K. established the Office for Standards in Education in 1992 as a national education supervision agency. South Africa developed the National Quality Framework (NQF), under which numerous quality standards and benchmarking mechanisms were established.[3] In contrast, China's standards for the basic public education service system are inadequate. It is mainly shown in the narrow scope of service targets, low guarantee standards, unclear expenditure responsibilities, lack of quality monitoring and evaluation mechanisms, and limited coverage.

Generally, in China's current public education financial system, local governments have been delegated too much administrative power and too little financial power. The local government bears preschool education spending, which inevitably affects its enthusiasm for education development. Second, the narrow coverage of service objects deviates from preschool education's basic attributes of inclusiveness and for public welfare. Third, the compulsory education spending responsibility is stipulated to be "shared proportionally by the Central and local governments," but the proportion is not clearly defined. Last but not least, the basic public education service system should be not only "compensatory" but also "inclusive," and should compensate vulnerable groups on the basis of equal opportunity.

The Plan for the Development of the National Basic Public Service System During the 12th Five-Year Plan Period sets clear national basic standards for China's basic public education services in the 12th Five-Year Plan period. It makes an overall arrangement of the service items, service

[3] Wang Su, Zou Junwei, Sun Yuze, Fang Yong, "International comparative analysis of public education services," in *China Education Policy Review*, ed. Yuan Zhenguo (Beijing: Educational Science Publishing House, 2011).

objects, support standards, expenditure responsibilities, and coverage of China's basic public education services.

Unreasonable Power and Responsibility Delegation Between the Central and Local Governments in Basic Public Education Services
The basic principle of expenditure responsibility and administrative power delegation between the central and local governments is that national or inter-provincial (autonomous regions or municipalities directly under the central government) affairs are decided, undertaken, and managed by the central government, while local affairs are decided and undertaken by local governments as approved by the central government's decrees, and managed by the local government. If the local government's revenue cannot cover the spending in some local affairs, the central government will, in principle, cover subsidies through general transfer payments. For the affairs entrusted by the central government to the local government, the central government will arrange the funds in full through special transfer payments. Regarding the common affairs of the central and local governments, the respective authority and spending ratio shall be determined on a case-by-case basis.

Table 3.10 shows the division of education spending between the central and local governments in OECD countries. On average, the central government of OECD countries undertakes 50.7% of education spending. Most of their education spending comes from the central or state governments, while county-level governments shoulder a lesser part. Comparatively, China's unreasonable division of education powers and responsibilities affects the enthusiasm of local governments to develop education undertakings and hinder the construction of an institutionalized basic public education service system.

In 2020, China's basic public education service system will need to form a power-responsibility delegation pattern where the central government is responsible for a larger part of expenditure responsibilities and local governments take administrative power. It is necessary to ensure that the provincial government executes the education planning power and vigorously promotes the balanced development of basic public education services.

Backward Legal System Construction for Basic Public Education Services
The basic public education service system is to a systematic and holistic arrangement comprised of the scope and standards, resource allocation, management and operation, supply methods, and performance evaluation

Table 3.10 Education spending responsibility division among governments at different levels in OECD countries in 2009 (%)

Country	*Initial funds of governments at different levels before transfer payments*				*Ultimate funds of governments at different levels after transfer payments*			
	Central government	*State government*	*Local government*	*Total*	*Central government*	*State government*	*Local government*	*Total*
Australia	38.7	61.3	0.0	100	3.3	96.7	0.0	100
Canada	3.5	73.7	22.8	100	2.8	11.1	86.1	100
France	69.2	19.0	11.9	100	69.0	19.0	12.0	100
Germany	10.3	72.2	17.5	100	8.5	68.5	23.0	100
Japan	18.0	64.2	17.8	100	0.7	81.5	17.8	100
U.K.	25.4	0.0	74.6	100	25.4	0.0	74.6	100
U.S.	10.0	38.7	51.3	100	0.4	1.6	98.0	100
Average	50.7	24.0	27.1	100	34.1	24.6	43.6	100

Source: Wang Jian, "Governance allocation and expenditure responsibility division of basic public service: A case of OECD countries," *Journal of University of Electronic Science and Technology of China (Social Sciences Edition)*, 04 (2013): 49–55

of basic public education services. At present, most countries' basic public education services are secured by sound legislative measures, as "law-based education governance" has become the consensus of the international community. Countries at different levels of economic development and even with different political systems all attach great importance to establishing a basic public education service system through legislation. They incorporate different types and levels of educational public services into the public service construction system through legislation, embedding the construction of basic public education services into laws and norms. The Third Plenary Session of the 18th CPC Central Committee proposed to encourage the private sector to run schools and improve government procurement of services, which put forward specific requirements for the construction of China's basic public education service system. China has not yet introduced a preschool education law, and the backward preschool education legislation has directly affected the process of incorporating preschool education into basic public education services. The Vocational Education Law of the People's Republic of China introduced in 1996 lacks provisions specific on how secondary vocational education and higher vocational education should be included in the basic public education service system, failing to meet the practical demands of the current construction of China's basic public education service system. China has neither introduced a law on government purchase of service. Such practices mainly observe relevant provisions in the Government Procurement Law of the People's Republic of China, which is mainly designed for the government's "internal services" and does not apply to "basic public services." Many practices, such as the inventories, prices, rights and responsibilities, fund management, and risk-taking of services purchased by the government, cannot find operational reference from provisions in the Government Procurement Law.

3.3 Long-Term Prospects and Development Basis of the Basic Public Education Service System

3.3.1 *Development Objectives of China's Basic Educational Public Service System in 2020*

Overall Outlook

In 2020, China will basically realize education modernization, form a learning society, and become a country with a quality workforce. The country will establish educational equity that benefits all and adheres to the public welfare and inclusiveness of education to ensure that citizens enjoy their legitimate rights to good education. There will be a sound basic public education service system. The gap between urban and rural areas in basic public services will be significantly narrowed, and basic public education services will be basically equalized.

Development Objectives and Standards

Development Objectives

Efforts will be put into consolidating and improving nine-year compulsory education and popularizing senior high school education and preschool education. The government-led, multi-approach financial aid policy for students from financially disadvantaged families will be improved, and the basic public education service system will be upgraded.

Nine-years compulsory education. (1) The achievements in compulsory education popularization will be consolidated. Compulsory education will be improved in every aspect. Development within a region will be basically balanced. School-aged children and adolescents will be ensured to receive good compulsory education. (2) Overall planning will be made for school layout and standardizing compulsory education schools. (3) IT-based instruction will be promoted in primary and secondary education. (4) Essential-qualities-oriented (EQO) education will be comprehensively implemented. Curriculum reform and instruction approach reform will be advanced. The basic standards and monitoring system of the national compulsory education quality will be improved. The academic burden of primary and secondary school students will be effectively reduced. (5) Compulsory education teachers' caliber will be enhanced, and there will be stronger bilingual teaching staff in ethnic-minority areas. (6) More public education resources will be allocated to rural areas, border areas, poverty-stricken areas, ethnic-minority areas, and old revolutionary base areas. A unified standard will be adopted for the staffing and salary of urban and rural primary and secondary school teachers within the county (city or district). An exchange system for teachers and principals will be implemented, and key schools and key classes will be eliminated in compulsory education. (7) Better boarding primary and secondary schools will be built in rural areas. Living allowances for boarding students from financially disadvantaged families in rural compulsory education will continue to rise. Living subsidies will be granted to all orphan boarding students in compulsory education. (8) A nutrition improvement plan will be implemented for rural compulsory education students. (9) The resource collaboration and sharing system and paired exchange and assistance system among urban and rural compulsory education schools will be improved to ensure that migrant workers' children living with their parents receive compulsory education on an equal basis. The entrance examination after compulsory education in the local area for them will be improved.

Senior high school education. (1) Senior high school education will be popularized, junior high school graduates' needs for senior high school education will be met, and the gross enrollment rate will be driven to 90%. (2) Support will be increased for senior high school education in poverty-stricken areas in the central and western China. (3) Diversified systems will be promoted for running general senior high schools, and high-quality resources will expand. (4) Diversified training models will be promoted to meet the development needs of students with different potentials. (5) General senior high schools will be promoted to develop with their own characteristics. (6) The state funding system for students from families with financial difficulties in general senior high schools will be improved.

Government coordination will be strengthened to promote the coordinated development of general senior high school and secondary vocational education. Secondary vocational education will be vigorously developed, which will be committed to services and oriented by employment. Equal efforts will be put in school education and vocational training. The industry-university cooperation mechanism will be improved. A vocational education training model of work-study integration, school-enterprise cooperation, and on-the-job internships will be in place. The direct admission system for vocational school graduates will be improved and the channels for graduates to continue their studies will be increased. The goal is to form a modern vocational education system that adapts to the transformation of economic development patterns and industrial structure adjustment, manifest the idea of lifelong education, and meets the socio-economic needs for high-caliber workers and skilled talents.

There will be stronger vocational education teaching staff, and skilled talents will be encouraged to teach at vocational schools. Basic capacity building will be strengthened in secondary vocational education. A policy of tuition-free secondary vocational education will be implemented for rural students, urban students from financially disadvantaged families, and students studying agriculture. The subsidy policy for students from financially disadvantaged families will be improved. Efforts will be put in introducing free secondary vocational education.

Preschool education. (1) One-year preschool education will be universally popularized, two-year preschool education will be basically popularized, and three-year preschool education will be popularized in areas where conditions permit. The goal is to achieve a gross enrollment rate of 95.0% for one-year preschool education, 80.0% for two-year, and 70.0% for three-year. (2) A government-led, socially engaged, public- and private-run kindergarten system will be established. A preschool educational public service system that covers urban and rural areas with a reasonable layout will be built. (3) The private sector will be encouraged to establish kindergartens, and private kindergartens will be actively supported. They will be guided and supported to provide inclusive services by means of government purchases, rent reduction or exemption, substituting awards for subsidies, and dispatching public preschool teachers. (4) Surplus school buildings and teachers of primary and secondary schools will be fully used for kindergartens (preschool classes). (5) Bilingual preschool education will be actively developed in ethnic-minority areas. (6) Financial support will be provided for children and orphans from families with financial difficulties to receive preschool education. (7) There will be stronger preschool teaching staff.

Special education. (1) In most parts of the country, each city (prefecture) and county (city) with a population of more than 300,000 and many children and adolescents with disabilities will have a special education school. (2) Schools at all levels and of all types will actively facilitate the admission of people with disabilities, and continuously expand LRCs and special education classes in general schools. (3) Compulsory education will be further popularized among children and adolescents with disabilities. Senior high school education for people with disabilities will be accelerated. Vocational education for people with disabilities will be vigorously promoted. Efforts will be put in developing higher education for people with disabilities. Preschool education will be developed for children with disabilities according to local conditions. (4) The feasibility of 12- to 15-year free education will be put under discussion.

Lifelong education. (1) A flexible and open lifelong education system and a lifelong learning "overpass" will be built. Public finance will be used to ensure a society where every citizen is committed to learning. (2) Various types of education at all levels will be connected. Multiple options will be provided to meet individuals' diverse learning and development needs.

(3) Multiple ways will be sought for government-provided education. The community will be the basis of practice. Pilot programs of community learning centers will be conducted.

Basic Standards

A national standard system for basic public education services will be established and improved according to relevant national laws and regulations to ensure the scale and quality of public education services, define the powers and expenditure responsibilities of different tasks, and balance the development in urban and rural areas.

Public funds for rural primary and secondary schools in compulsory education will be subject to a nationally unified benchmark. Standards for school building construction, equipment allocation, teacher appointment, and instruction management specifications will be developed and implemented by the Ministry of Education according to law with competent departments and the national standardization administration.

Provinces (autonomous regions and municipalities directly under the central government) will observe and implement basic national standards, and may expand the scope of basic public education services and raise the service standards based on the actual conditions of the region. Table 3.11 shows the standards that the basic public service system should meet in 2020.

3.3.2 Development Objectives and Basis of China's Basic Educational Public Service System in 2030

Overall Outlook

A state-led and socially engaged basic educational public service system will be established. There will be a sound basic educational public service system. The gap in the basic educational public service system between urban and rural areas and between regions will be significantly narrowed, and equalized basic public education services will be achieved. The quality of education will be improved, developing toward diversification and realizing a modernized basic educational public service system. The basic educational public service system will be expanded to cover preschool education, compulsory education, secondary vocational education, higher vocational education in some areas of national priority development, and special education. The legal system for the basic educational public service system will be improved, and support for vulnerable groups will be increased.

Table 3.11 Standards for the basic public education service system (in 2020)

Service	*Object*	*Standard*	*Expenditure responsibilities*	*Coverage*
Nine-year compulsory education				
Free compulsory education	School-aged children and adolescents	Tuition fees, miscellaneous fees, and accommodation fees for rural boarding students are exempted, and textbooks are provided to them for free. The average annual public funding per rural primary and secondary school student is not less than CN¥700[a] for general primary schools and CN¥980 for general junior high schools.	Shared proportionally between the central and local governments	100% covering the target group, with a completion rate of nine-year compulsory education of 95%
Living allowance for boarding students	Boarding students from rural families with financial difficulties	Annual allowance per primary school student: CN¥1400; per junior high school student: CN¥1750	The local government is responsible, with appropriate central financial subsidy.	100% covering the target group
Nutrition improvement for rural compulsory education students	Compulsory education for students in poverty-stricken rural areas	Apart from the living allowance for boarding students, students in special poverty-stricken areas will receive CN¥4.2 of nutrition allowance per student per day (calculated by 200 days in school in a year).	The local government is responsible. The central government will fund the national pilot areas and offer appropriate financial subsidy to other aeras.	100% covering the target group

(*continued*)

Table 3.11 (continued)

Service	*Object*	*Standard*	*Expenditure responsibilities*	*Coverage*
Senior high school education				
Free secondary vocational education	Rural students, students from urban families with financial difficulties, and students studying agriculture	Tuition exempted	Shared proportionally between the central and local governments	100% covering the target group, with a gross enrollment rate of senior high school education of 90%
National grant for secondary vocational education	Full-time rural students and students from urban families with financial difficulties	No less than CN¥2100 per student per year for two years	National grant for secondary vocational education	Full-time rural students and students from urban families with financial difficulties
National grant for general senior high school education	Students from families with financial difficulties	CN¥2100 per student per year in average. The local government may determine between CN¥1400 to CN¥4200 according to local conditions.	Shared proportionally between the central and local governments	100% covering the target group
Preschool education				
Preschool education grant	Children and orphans from families with financial difficulties	The specific funding approaches and standards are determined by the local government.	The local government is responsible, with appropriate central financial subsidy.	100% covering the target group, with a gross enrollment rate of one-year preschool education of 95%, two-year, 80%, and three-year, 70%

(*continued*)

Table 3.11 (continued)

Service	*Object*	*Standard*	*Expenditure responsibilities*	*Coverage*
Special education				
Special education grant	School-aged children and adolescents with disabilities	Free education for about 12 years is implemented according to regional differences. Other specific funding approaches and standards are determined by the local government.	Shared proportionally between the central and local governments	Compulsory education for the disabled basically popularized, with an enrollment rate of 93%, and 100% covering the target group
Lifelong education				
Lifelong education universally popularized	Adult workers, middle-aged and elderly people with educational needs	Establishing learning centers based on communities (neighborhood committees and villagers' committees)	For the basic content, shared between the central and local governments	Establishing pilot learning centers, seeking effective learning models, summarizing principles and standards for establishing such centers

[a]It is calculated at a constant price, and increases with the growth of GDP based on the 2015 funding, that is, funding = (1 + 0.07) ^ 5. According to Li Shantong's forecast, the average annual growth rate from 2016 to 2020 is expected to remain at about 7.0%, and it is expected to maintain a growth rate of about 6% before 2030

Development Objectives and Standards

Development Objectives

Preschool education. (1) A government-led, socially engaged, public- and private-run kindergarten system will be established. A preschool educational public service system that covers urban and rural areas with a reasonable layout will be built. (2) Two-year preschool education will be popularized and three-year preschool education will be basically popularized. The goal is to achieve a gross enrollment rate of 98% for one-year preschool education, 90% for two-year, and 85% for three-year. (3) Children and orphans from families with financial difficulties will receive free pre-school education. (4) Efforts will be put in the education and development for infants aged from 0 to 3.

Compulsory education. (1) The achievements of balanced development within the region will be consolidated. Education quality will be enhanced to the world's advanced level. (2) Better primary and secondary boarding schools will be built in rural areas. Dormitory conditions and safety standards will be improved. Dormitory management will be strengthened. (3) Free nutritious meals will be provided for eligible students.

Senior high school education. (1) Senior high school education will be popularized, with a gross enrollment rate of over 95%. (2) Diversified general senior high schools and school-running systems will be promoted with high-quality resources. (3) General education and vocational education will be further integrated to provide students with more learning opportunities and resources and form a unique educational style and school culture. (4) A supervision and evaluation system will be established and improved to promote the diversified development of general senior high schools.

Vocational education. (1) Secondary vocational education will be vigorously developed, which will be committed to services and oriented by employment. Equal efforts will be put in school education and vocational training. (2) The industry-university cooperation mechanism will be improved. A vocational education training model of work-study integration, school-enterprise cooperation, and on-the-job internships will be fully implemented. (3) Secondary vocational education will be provided for free.

Some national strategic priority areas will be included in the basic public service system through legislation. Higher vocational education in the national priority development areas will be strengthened, and vocational training will be expanded. The skilled personnel training system will be improved. Equal efforts will be put in curricula education and vocational training, and vocational schools and general schools will be encouraged to open up courses and teaching resources to each other. The vocational training function of higher vocational schools will be strengthened, and schools will be encouraged to carry out vocational training at multiple levels and in multiple forms for the society. Industries and enterprises will be encouraged to carry out various forms of vocational training.

Special education. (1) Each city (prefecture) and county (city) with a population of more than 300,000 and many children and adolescents with disabilities will have a special education school. (2) The development of preschool education for the disabled will be emphasized. Compulsory education for the disabled will be popularized, with a gross enrollment rate of over 95%. (3) A 12- to 15-year free education will be introduced. Vocational education for the disabled will be vigorously promoted. (4) General education and special education will be integrated, and schools at all levels and of all types will basically meet the conditions for accepting people with disabilities and will be equipped with professional teaching assistants.

Lifelong education. (1) The open lifelong education system will be improved and a lifelong learning "overpass" will be built. An effective supply model will be formed with public funding to ensure a society where every citizen is committed to learning. (2) The vertical connection and horizontal communication of various types of education at all levels will be promoted, community learning centers will be built, and multiple options will be provided to meet individuals' diverse learning and development needs. A lifelong learning culture will be established.

Basic Standards

Basic standards will be established according to national laws and regulations on education, to ensure the scale and quality of public education services, define the powers and expenditure responsibilities of different tasks, and promote balanced development between urban and rural areas. The stable supply of basic services will be guaranteed by law. The national standard system for basic public education services will be further improved.

Public funds for rural primary and secondary schools in compulsory education will be subject to a nationally unified benchmark. Specific standards for school building construction, equipment allocation, teacher allocation, and instruction management codes will be developed and implemented by the Ministry of Education according to law with competent departments and the national standardization administrative department.

Provinces (autonomous regions and municipalities directly under the central government) will observe and implement basic national standards, and may expand the scope of basic public education services and raise the service standards based on the actual conditions of the region. Table 3.12 shows the standards that the basic public service system shall meet in 2030.

Table 3.12 Standards for the basic public education service system (in 2030)

Service	*Object*	*Standard*	*Expenditure responsibilities*	*Coverage*
Nine-year compulsory education				
Free compulsory education	School-aged children and adolescents	Tuition fees, miscellaneous fees, and accommodation fees for rural boarding students are exempted, and textbooks are provided to them for free. The average annual public funding per rural primary and secondary school student is not less than CN¥938[a] for general primary schools and CN¥1313 for general junior high schools.	Shared proportionally between the central and local governments	100% covering the target group, with a completion rate of nine-year compulsory education of 98%
Living allowance for boarding students	Boarding students from rural families with financial difficulties	Annual allowance per primary school student: CN¥1876; per junior high school student: CN¥2345	The local government is responsible, with appropriate central financial subsidy	100% covering the target group (continue to consolidate and raise the standard)
Nutritional lunch for compulsory education students	Compulsory education for students in all areas	CN¥5.63 of nutrition allowance per student per day (calculated by 200 days in school in a year).	The local government is responsible, with appropriate central financial subsidy.	100% covering the target group (continue to consolidate and raise the standard)
Senior high school education				
Free secondary vocational education	All students	Tuition exempted	Shared proportionally between the central and local governments	100% covering the target group, with a gross enrollment rate of senior high school education of 95%

(*continued*)

Table 3.12 (continued)

Service	*Object*	*Standard*	*Expenditure responsibilities*	*Coverage*
National grant for secondary vocational education	Full-time rural students and students from urban families with financial difficulties	No less than CN¥2814 per student per year for two years	Shared proportionally between the central and local governments	100% covering the target group (continue to consolidate and raise the standard)
National grant for general senior high school education	Students from families with financial difficulties	CN¥2814 per student per year in average. The local government may determine between CN¥1876 to CN¥5628 according to local conditions	Shared proportionally between the central and local governments	100% covering the target group (continue to consolidate and raise the standard)
Preschool education				
Preschool education grant	Children and orphans from families with financial difficulties	The specific funding approaches and standards are determined by the local government	The local government is responsible, with appropriate central financial subsidy.	100% covering the target group, with a gross enrollment rate of one-year preschool education of 98%, two-year, 90%, and three-year, 85%
Special education				
Special education grant	School-aged children and adolescents with disabilities	Free education for about 12 years is implemented according to regional differences. Other specific funding approaches and standards are determined by the local government.	Shared proportionally between the central and local governments	100% covering the target group; compulsory education for the disabled popularized, with an enrollment rate of 95%; models diversified; basic integrated education policies introduced

(*continued*)

Table 3.12 (continued)

Service	*Object*	*Standard*	*Expenditure responsibilities*	*Coverage*
Lifelong education				
Lifelong education universally popularized	Adult workers, middle-aged and elderly people with educational needs	Establishing learning centers based on communities (neighborhood committees and villagers' committees)	For the basic content shared between the central and local governments	Establishing one to two learning centers in the jurisdiction of a neighborhood committee or villagers' committee according to actual conditions; relevant regulations introduced; a society where each citizen is committed to learning formed

[a]It is calculated at a constant price, and increases with the growth of GDP based on the 2020 funding, that is, funding = (1 + 0.06) ^ 5. According to Li Shantong's forecast, the growth rate is expected to maintain at about 6% before 2030

3.3.3 *Development Basis of the Basic Educational Public Service System*

(1) The fundamental nature of the basic educational public service system determines that it must be led by the state, while the private sector can be encouraged to engage.

It is the bounden duty of the state to provide basic education as public goods. The basic educational public service system is a public good with strong positive social externality. It must be led by the state to prevent market failure. On the other hand, the market competition mechanism must be introduced into the public utilities sector to overcome the public utilities sector's low efficiency and overstaffed institutions and to meet diverse educational needs.

Internationally, according to the data released by the OECD, government investment dominates in primary education, secondary education, and postsecondary education beyond higher education. The government is the main investor for national education, but there are still private investments (see Table 3.13), though in a very small proportion. For example, in primary education, secondary education, and postsecondary education beyond higher education in the U.S., government funds account for 3.9%, while private funds account for 0.3%.

With reference to international models and based on China's conditions, China's preschool education can form a government-led, socially engaged, and public- and private-run kindergarten system in 2030, so that focus can be placed on ensuring the investment in developing primary education, secondary education, and postsecondary education beyond higher education.

Private funding is significantly greater in higher education than in other education degrees. Private funding is greater than government funding in higher education in the U.S., the U.K., Australia, Japan, and South Korea. This further illustrates the nature of quasi-public goods of higher education, whose funding sources can be allocated by the market. Therefore, higher education can be excluded from the basic educational public service system.

(2) The level of economic development determines the breadth and depth of China's basic educational public service system.

The World Bank estimates that the annual growth rate of China's GDP will be 7.0% from 2016 to 2020, 5.9% from 2021 to 2025, and 5.0% from 2026 to 2030.[4] China has the potential to become a modern, harmonious, and creative high-income society by 2030. According to statistical analysis, China's per capita GDP will reach US$7000–8000 in 2020, China will reach the level of a moderately developed country in 2030. Of course, there are some uncertainties.

Among moderate developed countries, South Korea provides nine-year free compulsory education, and compulsory education in South Africa also lasts nine years. Among developed countries, Japan provides nine years of free compulsory education, the U.K. implements an 11-year free compulsory education system, and Germany provides 12 years of

[4] World Bank and Development Research Center of the State Council, the People's Republic of China, *China 2030: Building a Modern, Harmonious, and Creative High-Income Society* (Beijing: China Financial and Economic Publishing House, 2013).

Table 3.13 Percentage of education spending in GDP (by funding sources and education degrees, in 2009)

Country	*Preschool education*			*Primary education, secondary education, and postsecondary education beyond higher education*			*Higher education*			*Education at all degrees*		
	Government	*Private*	*Total*	*Government*	*Private*	*Total*	*Government*	*Private*	*Total*	*Government*	*Private*	*Total*
U.S.	0.34	0.08	0.43	3.9	0.3	4.3	1.0	1.6	2.6	5.3	2.1	7.3
U.K.	0.26	n	0.26	4.5	n	4.5	0.6	0.7	1.3	5.3	0.7	6.0
Sweden	0.73	n	0.73	4.2	n	4.2	1.6	0.2	1.8	6.6	0.2	6.7
Germany	0.44	0.19	0.63	2.9	0.4	3.3	1.1	0.2	1.3	4.5	0.8	5.3
Australia	0.06	0.05	0.11	3.6	0.6	4.2	0.7	0.9	1.6	4.5	1.5	6.0
Italy	0.46	0.04	0.51	3.3	0.1	3.4	0.8	0.2	1.0	4.5	0.4	4.9
Japan	0.10	0.12	0.22	2.7	0.3	3.0	0.5	1.0	1.6	3.6	1.7	5.2
South Korea	0.11	0.15	0.26	3.6	1.1	4.7	0.7	1.9	2.6	4.9	3.1	8.0
Russia	0.74	0.13	0.87	2.3	0.1	2.4	1.2	0.6	1.8	4.7	0.8	5.5
Argentina	0.46	0.21	0.66	4.4	0.5	4.9	1.1	0.3	1.4	6.0	1.0	7.0
New Zealand	0.48	0.10	0.58	4.5	0.7	5.2	1.1	0.5	1.6	6.1	1.3	7.4

Note: *n* means that the data value can be ignored in the calculation. Some education degrees in Japan include others compulsory education.[5] Among the BRICS countries, Brazil implements eight-year compulsory education, Russia implements 11-year compulsory education, and India implements eight-year compulsory education. Considering the level and speed of China's economic development, it is more appropriate to maintain the compulsory education period as nine years in China, and it is necessary to improve education quality and provide higher-quality educational resources.

According to Xinhuanet, the share of China's education spending in GDP has been growing (see Fig. 3.2), while the spending in primary, secondary, and postsecondary education in moderate developed countries such as South Korea, New Zealand, and Argentina in 2009 was more than 4% of their GDP, respectively, 4.7%, 4.9%, 5.2% (see Table 3.13). China's basic education has a weak foundation. The long-term insufficient investment has led to low education quality, backward instruction facilities, and enormous regional differences. Therefore, for long in the future, China's education spending needs to focus on rural and underdeveloped areas, and subsidies should be increased for students from financially disadvantaged families to ensure educational balance.

In terms of educational attainment, according to the data in the Human Development Report 2013 (see Table 3.14), 62.7% of the Chinese population aged 25 and above had received secondary and higher education by 2010, which was significantly lower than that in developed countries (94.5% in the U.S., and 99.7% in the U.K.) and moderately developed countries (85.4% in South Korea, and 94.7% in Argentina). In terms of gross enrollment rate, primary education in China has been basically popularized, which was almost at the same level of developed and moderately developed countries. The gross enrollment rate of secondary education in China was 81.0%, lagging behind developed and moderately developed countries. The gross enrollment rate of secondary education in the U.K., Sweden, Germany, Australia, Japan, and New Zealand was 100% or even higher, and the total enrollment rate of secondary education in Russia and Argentina was close to 90%. In 2010, China's gross enrollment rate of

[5] Jiang Yungen, "The outcome and enlightenment of public education service reform in other countries," in *China Education Policy Review*, ed. Yuan Zhenguo (Beijing: Educational Science Publishing House, 2011).

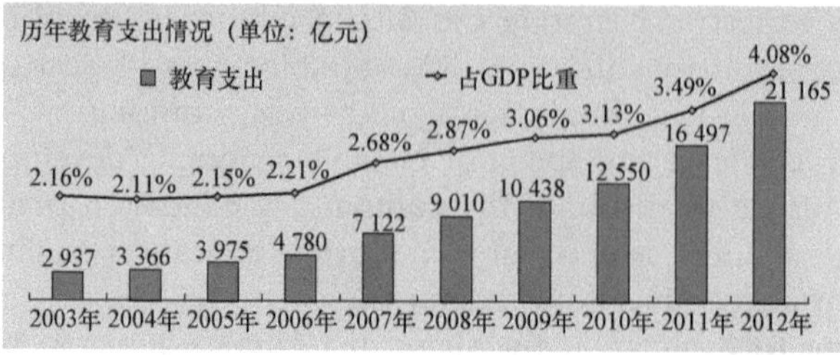

历年教育支出情况（单位：亿元）	Education spending through the years (unit: CN¥100 million)
教育支出	Education spending
占 GDP 比重	Share in GDP
YYYY 年	YYYY

Fig. 3.2 Share of education spending in GDP reaching 4% for the first time in 2012

senior high schools was 82.5%, an increase of 3.3% over 2009.[6] This figure reached 84.0% in 2011[7] and 85% in 2012.[8] The Outline of the National Plan for Medium and Long-Term Education Reform and Development (2010–2020) proposes that the gross enrollment rate of senior high schools should reach 87% in 2015 and 90% in 2020. Based on the speed of economic and education development, China's gross enrollment rate of senior high schools is expected to exceed 90% in 2020 and be over 95% in 2030. Twelve-year basic education (nine-year compulsory education and three-year senior high school or secondary vocational education) will be popularized.

(3) Education is a means to develop human resources and key to improving national competitiveness.

[6] http://www.edu.cn/jiao_yu_fa_zhan_498/20110708/t20110708_647259_1.shtml.
[7] http://www.edu.cn/jiao_yu_fa_zhan_498/20110708/t20110708_647259_1.shtml.
[8] http://www.chinanews.com/edu/2013/08-16/5173046.shtml.

Table 3.14 Educational attainment and gross enrollment rate at all degrees

Country	*Educational attainment*	*Gross enrollment rate (%)*		
	Percentage of people receiving secondary and higher education in the population aged 25 and above	*Primary education*	*Secondary education*	*Higher education*
	2010	*2002–2011*	*2002–2011*	*2002–2011*
U.S.	94.5	102.0	96.0	94.8
U.K.	99.7	106.0	102.0	58.5
Sweden	85.0	100.0	100.0	70.8
Germany	96.5	102.0	103.0	—
Australia	92.2	104.0	129.0	75.9
Italy	72.8	103.0	99.0	66.0
Japan	81.1[a]	103.0	102.0	59.0
South Korea	85.4[a]	104.0	97.0	103.9
Russia	56.0[a]	118.0	89.0	71.2
Argentina	94.7[a]	99.0	89.0	75.9
New Zealand	83.7	101.0	119.0	82.6
China	62.7[a]	111.0	81.0	25.9

Note: "Gross enrollment rate" is the gross enrollment rate for a given education degree (primary, secondary, or higher), regardless of age, which is presented as a percentage in the formal school-aged population at the same education degree

[a]Is the estimated value of researchers in 2010

Education is a means to develop human resources. Generally, economically developed countries are developed in human resources (see Table 3.15). China's HDI in 2012 was 0.699, ranking 101st in the world,[9] with an average schooling period of 7.5 years for adults. It indicates China is still underdeveloped in human resources. In 2030, China should be a developed country in human resources, and the average schooling period for the working population aged 25–64 will be more than 12 years.

Human capital accumulation and modern economic growth are almost twins (see Fig. 3.3). In global modernization, many cases and much data show that successful economic catch-ups are accompanied by successful human capital catch-ups. In modern economic growth, if the per capita GDP is low, human capital will first increase rapidly as the per capita GDP grows. With the further improvement of the per capita GDP, the growth

[9]http://money.163.com/13/0315/14/8Q11STIB00253G87.html.

Table 3.15 Human resources development

Country type	*Average schooling period for the working population aged 25–64*	*HDI*	*Aggregate production rate (by PPP)*	*Number of scientists and engineers per million people*	*Representative*
Developed countries in human resources	Over 12 years	Over 0.930	US$45,000	Over 2800	US, Canada, UK, Japan, Norway, Germany, and Australia
Moderate developed countries in human resources	10–12 years	Over 0.800	US$30,000	Over 1500	Korea, France, Italy, Singapore, and Austria
Underdeveloped countries in human resources	Less than 10 years	Below 0.800	Less than US$25,000	Less than 1500	Mexico, the Philippines, Brazil, and India

Source: The Research Team for the Report on China's Education and Human Resources, *Stride from a Country of Tremendous Population to a Country of Profound Human Resources* (Beijing: Higher Education Press, 2003): 101

rate of human capital will decline. From 2020 to 2030, China will be at a stage of rapid human capital accumulation.

National competitiveness is a comprehensive national strength, which consists of the country's social form, cultural tradition, education scale and quality, scientific research strength, innovation ability, and human resources in international competition under a certain system.[10] In the era of new economy, a country's technological and educational competitiveness determines its national competitiveness, and education contributes to the advancement of science and technology, the accumulation of high-quality human capital, and the overall improvement of national quality. The WEF' *Global Competitiveness Report* takes education as an indicator of the evaluation system. In the 2013–2014 Global Competitiveness Rankings, China ranked 29th but still has work to do in education quality

[10] Zhao Hongbin, "Educational competitiveness is the cornerstone of national competitiveness," *Education Science*, 4 (2008): 710.

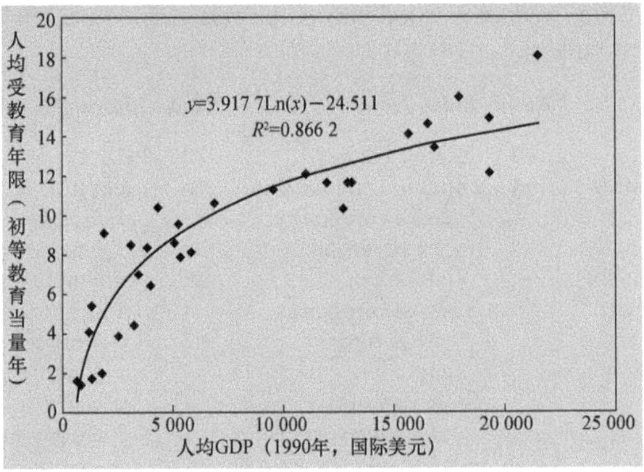

人均受教育年限（初等教育当量年）	Per capital schooling period (primary education equivalent year)
人均 GDP（1990 年，国际美元）	Per capital GDP (1990, in international dollar)

Fig. 3.3 The relationship between economic growth and human capital accumulation. (Note: This figure uses Angus Maddison's method to measure per capita schooling years in the primary education equivalent year, that is, weighting and converting the years of secondary education and higher education into years of primary education. The weights for years of secondary and higher education are 1.4 and 2.0, respectively. Source: The Research Team for the Report on China's Education and Human Resources, *Stride from a Country of Tremendous Population to a Country of Profound Human Resources* (Beijing: Higher Education Press, 2003): 17)

and innovation (see Table 3.16). The lack of innovation ability and labor education is significantly affecting China's economic growth.

(4) The transformation of national industrial structure demands high-caliber specialized personnel.

In 2010, the primary, secondary, and tertiary industries accounted for 10.2%, 46.9%, and 43.0%, respectively, in China's economy. Their contribution to GDP was 8.4%, 46.2%, and 45.4%, respectively. In 2009, the

Table 3.16 China's education and innovation-related indicators in WEF's global competitiveness rankings 2013–2014

Indicator	*Rank*	*Indicator*	*Rank*	*Indicator*	*Rank*
Pillar 4. Good health and primary education	40	Pillar 5. Higher education and training	70	Pillar 12. Innovation	32
1. Primary education quality	56	1. Gross enrollment rate of secondary education	90	1. Creativity	30
2. Net enrollment rate of primary education	4	2. Gross enrollment rate in tertiary education	83	2. Quality of scientific research institutions	41
		3. Quality of the education system	54	3. Scientists engineer availability	44
		4. Math and science quality	48	4. Valid patents	36
		5. School management quality	83		
		6. Availability of research and training services	62		

Source: Zhao Jinping, "Analysis and prospects of China's industrial structure change trends from 2010 to 2030," in *Research on China's Economic Growth Prospects and Population Aging*, ed. Zhao Jinping (Beijing: Economic Science Press, 2013)

actual employment distribution in the three industries were 38.1%, 27.8%, and 34.1%, respectively. In 2030, the secondary and tertiary industries will dominate, while the primary industry will decline (see Table 3.17).

In the future industrial structure transformation, it is necessary for China to accelerate the pace of innovation and establish an open innovation system. China needs to transform from labor-intensive to technology-intensive and provide high-quality export services through innovation. China also needs to improve resource utilization efficiency and seize "green" opportunities. The new energy, environmental protection, new energy vehicle, biotechnology, Internet of Things, and information industries will be emerging industries in China. All this demands China to popularize basic education and emphasize personalized training and carry out innovative education. According to the needs of industrial development, a basic balance within the region should be achieved in 2020. On this basis, higher-quality education should be provided and higher vocational education should be included in the basic public service system in 2030.

Table 3.17 Estimates for 2020 and 2030 in different settings (%)

Setting	*Year*	*Output Structure*			*Employment Structure*		
		Primary industry	*Secondary industry*	*Tertiary industry*	*Primary industry*	*Secondary industry*	*Tertiary industry*
Benchmark	2020	5.5	46.5	47.9	28.9	28.7	42.4
	2030	3.2	45.4	51.4	20.6	31.2	48.3
Model changed	2020	5.5	42.7	51.8	26.1	27.6	46.4
	2030	3.1	38.6	58.3	14.0	29.3	56.7

(5) Improving educational equity and providing high-quality education are important guarantees for the social harmony and economic growth.

Educational equity is the cornerstone of social equity. Equal educational opportunity allows reasonable class mobility by learning and hard working. Educational inequity will cause social instability and affect economic growth. According to the Outline of the National Plan for Medium and Long-Term Education Reform and Development (2010–2020), "Balanced development is a strategic task of compulsory education. We should establish and improve a mechanism for ensuring the balanced development of compulsory education. We should standardize compulsory education school construction, balance the allocation of teachers, facilities, books, school buildings, and other resources." "We should implement an exchange system for teachers and principals within the county (district)." "We should establish an urban-rural integrated compulsory education development mechanism, allocating more resources to rural areas in financial appropriations, school construction, and teacher staffing. We should achieve the balanced development between urban and rural areas first within the county (district), and promote it on a larger scale."

In practice, it is very difficult to implement the staff exchange and mobility system for principals and teachers. There are still great differences between provinces in teacher staffing, capital investment, and instruction facilities. Therefore, whether the balanced development of the basic educational public service system can be achieved in 2020 depends on the government's resolution and strength in implementation, whether there are sound support measures to encourage the mobility of teachers and principals, and whether there are sound measures to guarantee reasonable fund allocation in provinces.

In terms of special education, according to the Plan for the Development of China's Cause of People with Disabilities During 12h Five-Year Period, "Free compulsory education for students with disabilities will be fully implemented. For compulsory education students with disabilities, based on the policy of 'free textbooks, subsidies for lodging expenses and exemption from miscellaneous expenses,' subsidies will be further increased for their special needs. Free senior high school education for students with disabilities will be implemented in phases." In the international development of special education, the idea of inclusive education has prompted countries to integrate special education into general education to form a new educational trend that accommodates all students, opposes discrimination and exclusion, promotes student engagement, emphasizes students' collective cooperation, and meets the needs of different students.

3.4 Development Pathways of the Basic Public Education Service System

3.4.1 Promoting the Balanced Development of Basic Public Education Services

Enhancing Basic Public Education Services in Rural Areas

The basic educational public services in rural areas are mostly provided by local governments at all levels. Therefore, local governments should take the initiative to develop the economy and effectively improve their economic strength. This is the key to helping rural areas improve basic public education services. To this end, the central government should increase the special transfer payment for basic public education services in rural areas. When setting standard fiscal revenues and expenditures, decision-makers should consider the differences between rural areas in basic public service costs and ensure that the transfer payment is science-based and reasonable, and stabilize the supply of high-quality education resources for rural and remote areas.

Reasonably Allocating Education Resources

First of all, we should reform the staffing system for rural primary and secondary schools, improve the quality of the teaching staff, and establish a dynamic at-will employment system for rural teacher management. We should improve the teacher exchange system and promote the rational flow of teachers in the region. On the one hand, we may encourage teachers to rural areas with teacher subsidy and allowance policies. On the other hand, we may prioritize teaching at rural primary and secondary schools as a condition in the evaluation of teachers' professional titles. We should make full use of the high-quality teaching and information resources of urban schools in science-based education planning, and make them benefit a wider range.

Vigorously Developing Vocational Education

Against the backdrop of economic globalization and internationalized and popularized higher education, China should strengthen its human resources capabilities and establish first-class secondary vocational consortia, chains, and groups. We may use education vouchers, student loans, grants, and scholarships to provide education assistance to students from rural and urban low-income families, to create favorable conditions for the reform and development of higher vocational education.

Prioritizing Education in Ethnic-Minority Areas

Education development in ethnic-minority areas is a hallmark of educational equity. Since the founding of the People's Republic of China, especially the beginning of Reform and Opening Up, the CPC Central Committee and the State Council have attached great importance to education in ethnic-minority areas and implemented many preferential policies. We should improve the overall quality of teachers in ethnic-minority areas, strengthen teachers' morality and integrity, and insist on strict management of schools and teachers. We should vigorously promote information technology education and seize the "commanding heights" of education. Online learning can narrow the differences between ethnic-minority areas and developed areas, ensure the adequacy and pertinence of education resources in ethnic-minority areas, and lay a solid foundation for cultivating talents urgently needed in modern times.

Paying Attention to Education for Vulnerable Groups

Education in West China also deserves attention. Besides increasing transfer payments to the region, we may consider allocating a portion of the funds directly to county-level governments, especially those of poverty-stricken counties, and investing more of the newly added general fiscal transfer payments in education. We should pay attention to the education for migrant workers' children living with their parents, provide them with basic preschool education and compulsory education, and implement the college matriculation policy for migrant children. For left-behind children, teachers should pay attention to their mental health, provide psychological counseling, and guide students to establish proper outlook on society, on life, and on values. In this way, we will gradually expand the recipient scope of basic public education services.

3.4.2 Improving the Quality Assurance System for Basic Public Education

Improving the Accountability System for Local Governments
Accountability is an institutional design and arrangement. The accountability system for local governments prescribes government responsibility and legal responsibility. The legal responsibility that the local government shall bear is the administrative legal responsibility. The Regulation on the Punishment of Civil Servants of Administrative Organs and the provisions in educational laws and regulations on administrative legal responsibility provide normative basis for educational accountability. In the implementation of the accountability system, it is necessary to combine accountability with rewards and promotions to highlight the motivational effect of the system, which may serve as the basis for officials' assessment and promotion.

Establishing a Third-Party-Led Educational Evaluation and Education Service Survey System
Third-party evaluation can avoid the mandatory effect of the government in evaluation, better ensure a fair and impartial process, and democratize decision-making. This is a global trend in education. At present, the quality of China's basic public education services makes it urgent to establish diversified educational evaluation system led by third-party evaluation and accept social supervision. We should carry out diversified evaluations in terms of evaluation subjects, objects, models, and standards; reshape the evaluation team by engaging more parties; explore the construction of basic public education service think tanks; and draw lessons from all sectors.

TQM- and ISO9000-Based Instruction Quality Management
At present, China's basic public education services urgently needs quality improvement. The total quality management (TQM) and ISO9000 quality standards are useful for improving schools' instruction quality. In

market economy, basic public education services are public welfare undertakings and also service trade. As a special service industry, teaching can be regarded as a customer-service relationship. "Focusing on customers" is the primary principle of ISO9000, which is consistent with the proposition of China's basic education reform and the management rules of primary and secondary schools.

3.4.3 *Increasing the Efficiency of Basic Public Education Services*

Mindset Changing from Management to Governance
The government should change from a powerful manager to a service provider; create an easing, free, and autonomous environment; expand schools' autonomy in operation; and promote public and social engagement in management. The transformation of government functions can effectively address abuse, absence, or misplacement of regulation in public education services. The government should actively seek cooperation with NGOs, which can serve as a communication and buffer between the higher education community and the government.

Government Adjusting the Appropriation Approach
As the public's demand for basic public education services increases, the government should shift from direct granting to a combination of direct and indirect grants. Besides directly providing education services, the government can also support NGOs as education service providers. To measure the effect of education, the government's financial input should be linked with the output and effect, quality indicators should be added to the appropriation, and methods and standards should be introduced for performance appropriation.

Introducing a Fair-Competition and Free-Choice Mechanism
Introducing the fair-competition and free-choice mechanism of the market economy into education can change the government's monopoly on education and diversify education services. In Mexico and Brazil, the government meets the educational needs of low- and middle-income families and ensures educational equity with public education. Private schools target middle- and high-income groups and meet their selective needs. It is a new form of school choice for parents.

Establishing a Government-Led, Multiparty-Engaged School-Running System
Single education supplies can no longer meet people's needs. It is necessary to establish a diversified school-running system to provide people with a variety of basic public education services. On the premise of ensuring public education quality, we should develop private education, respect and meet people's diversified needs, and form a reasonable layout in which public and private schools coexist. We should define the scope and level of private capital in education, rationalize property rights issues, support the private sector in running schools in various forms, and encourage private capital to invest in education.

3.4.4 Upgrading the Legal System for Basic Public Education Services

Revising the Law on Government Purchase of Service
Government purchase of service mainly observes the Government Procurement Law of the People's Republic of China. However, many practices, such as the inventories, prices, rights and responsibilities, fund management, and risk-taking of services purchased by the government, cannot find operational reference from provisions in the Law. Therefore, it should be revised, and relevant content should be added. The service inventory should include public education planning and policy research, communication services, public education information collection and statistical analysis, public education infrastructure management and maintenance, public education result assessment, public education result exchange and promotion, and other education services.

Introducing and Improving Laws on Basic Public Education Services
The legislative lag on preschool education has affected its integration into basic public education services. A Preschool Education Law should be introduced as soon as possible to ensure the healthy and science-based development of preschool education. On the other hand, the Vocational Education Law should explain how to incorporate secondary vocational education and higher vocational education into basic public education services, so that vocational education can be better integrated with other public services such as culture, sports, and medical care. In this way, vocational education will lead the construction of the basic public service

system, promote the construction of a visionary and leading modern vocational education system, and realize the national strategic goal of building a country with a quality workforce.

Accelerating Legislation on Education Spending
Stable and adequate funding is the foundation for the provision of basic public education services. The Outline of the National Medium- and Long-Term Program for Education Reform and Development and the Opinions of the State Council on the Current Development of Preschool Education provide policy basis for the legislation on preschool education spending. However, the pace of legislation should be accelerated. The irrational division of compulsory education funding between governments at all levels has been a problem in many countries. We need to strengthen cultural exchanges with developed countries and draw lessons from them, to make theoretical preparations for the legislation on compulsory education spending.

Strengthening Law-Based Administration
In modern society, the government's education functions are mainly about providing basic educational public services and effective administration. For education, it is necessary to improve the legal mechanism for public education spending, increase investment in education according to law, and fully implement input responsibilities stipulated by the Compulsory Education Law, the Vocation Education Law, the Higher Education Law, the Teachers Law, and the Non-state Education Promotion Law. This is the basic premise to ensure the priority and scientific development of education. Strengthening law-based administration for education means implementing administration for education more often in ways such as legislation, planning, appropriation, and policy support under the principle of a "law-based, responsible, limited, and efficient government," so as to create a favorable environment for "law-based administration for education."

Establishing an Effective Supervision Mechanism
Basic public education services are compulsory and inclusive. Being compulsory means it must be provided by the government. Therefore, we must develop necessary supervision means and strictly prevent the worst scenarios. We should raise the basic standard to improve education quality and ultimately achieve the popularization of high-quality educational

resources and the general improvement of national quality. We should establish a supervision and inspection mechanism for basic public education services, ensure the implementation of national education policies, and protect the legitimate rights and interests of students, teachers, principals, and school-runners according to the law.

Law-Based Administration of Schools
Education disputes must be handled according to law. We should establish and improve education dispute resolution mechanisms and institutions, and urge and guide schools at all levels to establish and improve for complaint institutions and mechanisms students and teachers according to law. We should enhance the communication and education of rule of law and train specialized legal professionals for education. We should tighten the basic admission system for teachers and school managers, incorporate the assessment of educational laws and regulations into national examinations and qualification systems, and enhance teachers and school managers' rule-of-law awareness.

We should institutionalize and regulate school operation to enhance the school's overall rule-of-law awareness and form a healthy and orderly rule-of-law environment on campus. We should enhance rule of law in educational legislation, administration, and judiciary through institutional innovation to ensure the provision of high-quality educational services. We should maintain the harmony of the educational environment through the development of rule of law.

CHAPTER 4

Modernization of Basic Medical Care and Health Services

4.1 Significance of Enhancing Public Health and Fitness

Health is essential to an individual's growth and high-quality life. It is also a hallmark of a country's prosperity and people's happiness. Health enhancement and protection are integral to the well-being of all and continuous economic and social development. As the Declaration of Alma-Ata states, the attainment of health for all contributes to a better quality of life and to world peace.

For China, prosperity for all is impossible without health for all. With the improvement of China's comprehensive national strength and people's living standards, the people have more prominent and urgent health demands. Enhancing public health and fitness and providing equal basic medical care and health services for all citizens are conducive to social equity and justice, so that the fruits of development can be shared by the people. This is an important manifestation of CPC's belief of "putting people first and exercising governance for the people" and a necessary condition to achieve the two centenary goals.

K. Dong, N. Wei, *Transforming China's Public Services*,
https://doi.org/10.1007/978-981-99-3943-5_4

4.2 Defining Basic Medical Care and Health Services

Health services are a major issue concerning people's livelihood. The World Health Assembly 2005 stated in its resolution that everyone shall have access to health services without suffering from financial hardship as a result of paying for them. There are many ways to promote and maintain health, which may be impacted by education, housing, food, and employment. Addressing inequities in these areas can mitigate health inequities. Timely access to basic health services, such as health enhancement, disease prevention, and treatment and rehabilitation, is of extraordinary significance. Basic medical care and health services have a direct impact on people's rights to subsistence and development, social equity and stability, and economic and social development.

Basic medical care and health services are an important part of a country's public services and are sometimes called "basic public medical care and health services." "Universal coverage of basic medical and health care services" is the strategic objective of the deepening reform of medicine and healthcare system and the Chinese society's consensus. "Universal coverage" is essentially "equitable access," that is, any citizen, regardless of age, gender, occupation, region, or ability to pay, has the same rights. Although China has not yet made a universal definition of basic medical care and health services, it is generally believed that such a definition should have the following features: (1) Basic medical care and health services are those medical care and health services in the most urgent need. (2) They are provided to protect people's essential right to health. (3) They adapt to the current level of economic development and are affordable to the public sector. (4) They are not aimed at earning excess profits but should follow the cost-benefit principle.

China's present basic medical care and health services consists mainly of two parts. One is public health services, including disease prevention and control, planned immunization, health education, health supervision, maternal and child healthcare, mental health, health emergency, first aid, blood collection, food safety, occupational disease prevention, and safe drinking water. The other is basic medical care, including the diagnosis, treatment, and rehabilitation of acute and chronic diseases using basic medicines and appropriate technologies in accordance through standardized diagnosis and treatment procedures. Costs of basic medical care and health services are divided reasonably among the government, society, and the individual. Government funds support both the supply and demand sides of medical services, focusing on public health, rural health, urban community health, and basic medical security for urban and rural residents.

The Plan for the Development of the National Basic Public Service System During the 12th Five-Year Plan Period stipulates that during the 12th Five-Year Plan period, the Chinese government will:

(1) Provide urban and rural residents with national basic public health services, including free residents' health records, health education, vaccination, infectious disease prevention and treatment, maternal and child healthcare, elderly care, the management of hypertension and other chronic diseases, severe mental illness management, and health supervision and management; (2) carry out major public health programs, including the national immunization plan, prevention and control of major infectious diseases such as AIDS, tuberculosis, and schistosomiasis, subsidies for rural women's delivery in hospital, and cervical and breast cancer screening for women at the age; (3) implement the national essential drugs system, with all essential drugs included in the catalog of medicines covered by national medical insurance system and sold for no profit; and (4) provide assurance for medication safety among the public and ensure the quality and safety of drugs.

Different from basic medical care and health services, basic public health services are provided by the Chinese government to all residents for free, especially children, pregnant women, the elderly, and patients with chronic diseases, for the major health problems among urban and rural residents. They are an important part of basic medical care and health services.

A country's basic medical care and health services should be continuously adjusted and renewed according to economic and social development, changes in residents' health needs, and government affordability. As resources are always limited, the government should concentrate them in residents' most urgent health issues and ensure medical and health resources are optimally allocated.

4.3 Status Quo of and Problems and Challenges in Basic Medical Care and Health Services

4.3.1 *Status Quo*

China's Current Health Status

In the twenty-first century, China's health undertakings see significant progress, and urban and rural residents are healthier, which is mainly reflected in the three important indicators that are internationally recognized to comprehensively reflect health status. The first is life expectancy. Results of the sixth national census show that the average life expectancy of China's population in 2010 was 74.83 years, an increase of 3.43 years over that in 2000. By gender, the average life expectancy of males was 72.38 years, an increase of 2.75 years from that in 2000; females, 77.37 years, an increase of 4.04 years. At the same time, the average life expectancy of the world's population in 2010 was 69.6 years. The second is infant mortality, which dropped from 19‰ in 2005 to 10.3‰ in 2012. The third is maternal mortality, which dropped from 47.7/100,000 in 2005 to 24.5/100,000 in 2012. Although China is still a low- to middle-income developing country, the above-mentioned health indicators are already at the forefront of developing countries and reach the average of middle- and high-income countries.

Health Resources

China's total health expenditure in 2013 was CN¥3166.895 billion, including government health expenditure of CN¥954.581 billion (30.14%), social health expenditure of CN¥1139.379 billion (35.98%), and individuals' out-of-pocket (OOP) expenses of CN¥1072.934 billion (33.88%). Urban areas accounted for CN¥2364.495 or 74.7% of the total, while rural areas accounted for CN¥802.4 billion or 25.3%. The per capita health expense was CN¥2327.37, which was CN¥3234.12 in urban areas and CN¥1274.44 in rural areas. The total health expenditure occupied 5.57% of GDP. In 2014, China's total health expenditure was CN¥3531.24 billion, an increase of CN¥364.345 billion (11.5%) over that in 2013, and the per capita health expense was CN¥2581.66. The proportion of individuals' OOP expenses in the total health expenditure dropped from 33.88% in 2013 to 31.99% in 2014.

By the end of 2014, the total number of health care and medical institutions nationwide was 981,432, a y-o-y increase of 7034. They included 25,860 hospitals, 917,335 grassroots healthcare and medical institutions, 35,029 specialized public healthcare institutions, and 3208 other institutions. Grassroots healthcare and medical institutions included 34,238 community healthcare centers (clinics), 36,902 township health centers, 645,470 village clinics, and 200,130 clinics (medical rooms). Hospitals included 13,314 public hospitals and 12,546 private hospitals. In 2014, the number of private hospitals increased by 1233, and the proportion of private hospital beds in total hospital beds increased from 15.6% in 2013 to 16.8% in 2014. The growth rate of private hospital beds was 17.1%, higher than that of public hospitals of 6.7%. In general, the construction of the medical care and health service system has been significantly accelerated, and grassroots healthcare and medical institutions' service capabilities have been improved in all aspects.

By the end of 2014, the total number of health personnel nationwide was 10.2343 million, an increase of 443,700 (4.53%) over the previous year. They included 7.5898 million health technical personnel, 1,058,200 village doctors and health personnel, 379,700 other technical personnel, 451,300 administrative staff, and 755,300 logistic technical workers. The health technical personnel included 2,892,500 licensed physicians and physician assistants and 3,004,100 registered nurses. Compared with 2013, the health technical personnel increased by 379,200 people, or 5.26%. The urban and rural grassroots health service network has been strengthened through grassroots training programs and rural-urban integration. By the end of 2014, community healthcare centers had 488,771 staff, township health centers had 1,247,299 staff, and village clinics had 1,460,389 staff. The number of staff at community healthcare centers (clinics), township health centers, and village clinics across the country increased by 13,000, 13,000, and 3000, respectively, compared with those in 2013.

By the end of 2014, the number of physicians and physician assistants per 1000 population was 2.1; registered nurses per 1000 population, 2.2; and the total number of beds, 6.6012 million. The number of health institution beds per 1000 population increased from 4.55 in 2013 to 4.85 in 2014. Both the health workforce and beds grew faster than the population, and the number of nurses grew faster than that of physicians.

Medical Care and Health Services

In 2014, the total number of visits to medical and health institutions nationwide was 7.602 billion, an increase of 290 million (4%) over 2013. The per capita number of visits to medical and health institutions was 5.6, up from 5.1 in 2012. Among the total visits in 2014, 2.972 billion were to hospitals (about 39%), 4.364 billion to grassroots healthcare and medical institutions (about 57%), and 266 million to other medical institutions (about 4%). Compared with the previous year, the number of visits to hospitals increased by 230 million and that to grassroots healthcare and medical institutions increased by 40 million. In 2014, the number of visits to public hospitals was 2.65 billion (89.2% of the total), and that to private hospitals, 320 million (10.8%).

In 2014, the number of visits to township health centers and community healthcare centers (clinics) was 1.714 billion, which accounted for 22.55% of the total number of outpatient visits, a decrease of 0.2 percentage points (pps) from 2013.

In 2014, the number of hospital admissions in medical and health institutions nationwide was 204,411,800, an increase of 12,257,200 (6.38%) over 2013. Among them, the number of hospital admissions was 153,751,400 (about 75%), that to grassroots healthcare and medical institutions was 40,939,700 (about 20%), and that to other medical institutions was 9,720,700 (about 5%). Compared with the previous year, the number of hospital admissions increased by 13,677,100, that to grassroots healthcare and medical institutions decreased by 2.065 million, and that to other medical institutions increased by 645,100. In 2014, the number of public hospital admissions was 134.15 million (87.3%), and that to private hospitals was 19.6 million (12.7%).

Disease Prevention and Control

In 2014, a total of 3.076 million Class A and B infectious disease cases and 16,059 deaths from them were reported nationwide, with an incidence rate of 227.0/100,000 and a mortality rate of 1.2/100,000. The top five reported Class A and B infectious disease cases were viral hepatitis, tuberculosis, syphilis, bacterial amoebic dysentery, and gonorrhea, accounting for 90.4% of the total. The five most deadly Class A and B infectious diseases in reports were AIDS, tuberculosis, rabies, viral hepatitis, and human infection with H7N9 avian influenza, accounting for 98.2% of the total.

As the Patriotic Public Health Campaign continues to roll out, sanitation in urban and rural areas continues to improve. By the end of 2014, cumulatively 920 million rural residents in China had access to improved drinking-water sources, accounting for 95.8% of the total rural population; 79.0% of rural population had access to tap-water, an increase of 2.6 pps over that in 2013. A total of 199.393 million rural households had access to sanitary toilets, including an addition of 6.851 million that year. The coverage of sanitary toilets in rural areas was 76.1%, an increase of 2.0 pps over that in 2013. National basic public health service items and special programs for major public health services have been fully implemented, and the equalization of basic public health services has been further improved.

Maternal and Child Health

The increase in the average life expectancy in China is the combined result of the decline in mortality at all ages, especially the decline in infant mortality. In 2014, the prenatal examination rate of pregnant and lying-in women was 96.2%, and the postpartum visiting rate was 93.9%, which were both higher than those in the previous year. In 2014, the percentage of hospitalized delivery was 99.6% (99.9% in cities and 99.4% in counties), 0.1 pps higher than in 2013 (with that in cities the same as that in the previous year, while that in counties increasing by 0.2 pps). In 2014, the systematic management rate of children under the age of three was 89.8%, an increase of 0.8 pps over the previous year; that of pregnant women was 90.0%, an increase of 0.5 pps.

China's maternal mortality rate dropped from 23.2/100,000 in 2013 to 21.7/100,000 in 2014, achieving the target of the medical and healthcare system reform in the 12th Five-Year Plan period (to below 22/100,000). The infant mortality rate dropped from 9.5‰ to 8.9‰, achieving the target of the medical and healthcare system reform in the 12th Five-Year Plan period ahead of schedule (to below 12‰). According to the 2014 statistics, the mortality rate of children under the age of five was 11.7‰.

With the key factors affecting maternal and child health addressed, women and children have been healthier. In 2014, the prevalence of moderate to severe malnutrition among children under the age of five was 1.48%, down by 52.1% from that in 2000.

Healthcare Security

The basic healthcare security system has been improved. The coverage of the three basic medical insurances, namely the medical insurance for employees, the urban resident medical insurance, and the new rural cooperative medical care system, is stable at over 95%. By the end of 2014, 1.334 billion urban and rural residents were covered by basic medical insurances, including 283.25 million urban employees and 314.49 million urban residents. A total of 736 million people, or 98.9%, were covered by the national new rural cooperative medical care system. The per capita contribution has increased from CN¥42 at the end of the Tenth Five-Year Plan to CN¥411 in 2014, which marked significant improvement of security.

4.3.2 *Problems and Challenges*

Major Problems

Insufficient Total Health Expenditure
China's total health expenditure accounts for a small share in GDP. With public financing being low and government input being small, individuals' OOP health expenses are relatively high. Government budgetary health expenditure is also at a small proportion. In 2011, China's public financing accounted for 55.89% of the total health expenditure and 2.88% of the GDP, which failed to meet the standards of the World Health Organization (WHO). Compared with countries at different economic development levels, the share of public financing in China's total health expenditure in 2010 (54.31%) was only higher than the average of low-income countries (41.84%). Correspondingly, the share of residents' OOP health expense in cash (35.29%) was higher than that of the middle- and high-income country group (29.92%), where China is located, as well as the global average (32.09%).

A Significant Gap Between the Provision of Grassroots Medical Care and Health Services and the People's Health Needs
At present, China's medical resources per 1000 population is far less than that in developed countries, and the number of nurses per 1000 population is also far less than that in Brazil, which is also a developing country. Medical resources have an unbalanced supply structure and, with unreasonable allocation, are concentrated in the Grade-III Level-A hospitals in big cities. There is a serious imbalance in the allocation of medical resources between large hospitals and grassroots hospitals. Official data show 80% of China's medical resources concentrated in big cities, and 30% of them in big hospitals.

China's grassroots medical care and health system with rural grassroots and urban community medical institutions as the core fails to meet the people's growing demand for medical care and health services. The system is mainly undermined by the following problems: The first is inadequate, unstable, and insufficient health expenditure. There is significant variance between urban and rural areas. Second, the structure of grassroots medical care and health personnel is unreasonable. The workers' capacity needs to be improved, and there is a shortage of general medical personnel. Third, private medical care and health institutions fail to effectively supplement the grassroots medical care and health service system. Fourth, community health services' basic functions are not in place. There are also many problems in rural medical care and health service networks: The first is the absence of township and village networks, lack of grassroots health resources, poor medical conditions in village clinics, weak prevention and protection capabilities, and seriously insufficient funding at the county and township levels. Second, the needs of rural medical care and health services cannot be met due to the lack of medical care and health professionals and technical personnel, resulting in a slow overall technical improvement in rural diagnosis and treatment and that many diseases cannot be treated in rural areas. Third, networks at the three levels lack effective vertical connections. There are few effective connections between the county, town, and village levels, resulting in inefficient resource allocation.

Insufficient Capacity to Handle Major Public Health Emergencies

China has successfully dealt with many major public health emergencies in its long-term exploration and accumulated an abundance of theoretical and practical experience. However, there are still problems and deficiencies, mainly including the following. First, the prevention and monitoring system is weak. Second, government investment is insufficient; the central and local expenditure structure is unreasonable; the public health system is weak in both software and hardware; and the emergency management ability for public health emergencies is poor. Third, there lacks a comprehensive emergency command and decision-making mechanism; social mobilization for emergencies is weak; and the public is not engaged enough. Fourth, the legal system for responding to public health emergencies is defective, and health law enforcement supervision is not in place.

Shortage of High-Caliber Health Technical Personnel and Pronounced Structural Issues

Although the total number of health personnel in China seems very large, the number of physicians and nurses (nurse practitioners) per 1000 population in China is far behind that in developed countries. Moreover, the distribution variance between urban and rural areas is still prominent. The number of health technical personnel per 1000 population in urban areas was 2.2 times that of rural areas in 2005, 2.5 times in 2010, and 2.6 times in 2014. In 2014, there were 3.5 licensed (assistant) physicians, 4.3 registered nurses, and 9.7 health technical personnel per 1000 population in urban areas, and 1.5, 1.3, and 3.8 in rural areas, respectively. The difference in the number of licensed (assistant) physicians per 1000 population between urban and rural areas increased from 1.9 in 2013 to 2; that in registered nurses, from 2.8 to 3; and that in health technical personnel, from 5.6 to 5.9. The educational degree and professional title of most grassroots health personnel are low. The workforce structure is unreasonable. Workers' professional and technical capacity needs to be improved. Especially among rural grassroots health personnel, most of them—both licensed doctors and registered nurses—have only a technical secondary school or junior college degree, which seriously affects the level and quality of grassroots medical services.

Defective Development of the Medical Security System

China's current medical insurance system has the following major problems: The first is the limited coverage. There are too many restrictions on the scope of medical expense reimbursement, and the ratio and total amount of reimbursement are low, resulting in a heavy burden of medical expenses on people for serious illnesses. The second is the unified implementation of the insurance system. Under China's medical insurance system, there are differences between urban employees and urban residents, between rural residents and urban residents, between civil servants and employees of public institutions and employees of urban enterprises. There is also significant variance in the payment and claim standards between regions. Third, the pooled account of medical insurance applies limitedly, and the pooled account of the new rural cooperative medical care system applies only to the local county. The fourth is ineffective administration. The administration of medical insurance is decentralized in governments at different levels and in different departments. The medical insurance for urban employees and residents is administered by the social security department, while the new rural cooperative medical care system is administered by the health department. In terms of the payment and insurance compensations of rural medical expenses, public services are less commercialized; the personnel are mixed in qualifications and not professional enough, with seriously insufficient service capacity. The fifth is insufficient applicability of medical insurance. When seeking medical and health services in places other than the registered residence, people have to pay for the expensive services entirely on their own.

The System and Mechanism for the Development of Medical Care and Health Services Being Severely Problematic

The development of health services, restricted by increasingly prominent problems in the system and mechanism, has been unbalanced, uncoordinated, and unsustainable. For example, the reform of public hospitals still lacks consensus on the direction and approach, and some key issues need to be defined. Many public hospitals are built, renovated, or expanded at an excessive scale and over the standard. Public hospitals' excessive expansion and borrowing for construction will inevitably increase the pressure on hospital operations, resulting in repeated examinations and excessive medical treatment. It will further drain social capital for medicine and hinder a science-based and rational hierarchical medical system. On the other hand, as private medical institutions are underdeveloped, it is difficult for them to compete on an equal footing with public hospitals, thus forming a pattern of public hospital monopoly.

There are still major problems in the compensation mechanism for public hospitals. The prices of some medical services have not been adjusted for years. The prices of medical personnel's technical services are low, while the prices of medicines, consumables, examinations by large equipment, and some inspection items are inflated. Added with the combined effect of item-based pricing and payment, medicines are charged higher to make up for the low-priced medical services, distorting medical services. Most evaluation standards still overemphasize economic benefits and neglect social benefits. Instead of doctors' services, the income of public hospitals is mainly from medical examinations and medicines, which drives up their prices. The division of responsibilities and cooperation between urban hospitals and community health institutions in China is immature. The allocation of medical resources between large hospitals and grassroots hospitals is seriously unbalanced. Resources are distributed unevenly, showing an obvious trend of upward concentration. Medical services are severely segmented, with low overall service efficiency. Medical institutions are poorly connected and coordinated. The functions, responsibilities, and structure of urban medical institutions at all levels need to be optimized. There has not been a medical service model in which acute, difficult, critical, and severe diseases are treated at large hospitals, while minor and chronic diseases are treated in the community. Therefore, it is impossible to highly and completely integrate urban medical and health resources, not to mention improving the city's health service network. In addition, problems in public hospitals such as the obscure boundary between supervision and operation, between public welfare and profiting, are still to be solved.

Defective Drug Policies and Drug Production and Circulation Systems

Although China has policies and regulations on drug research, production, circulation, and use, due to the different perspectives of departments and local governments, some policies and regulations contradict in their purposes, and some procedures are difficult to be connected and coordinated. China so far does not have a comprehensive national drug policy.

The essential drugs system needs further studies and modification. For the essential drugs list, the basis of national drug policies, the health authorities and the labor and social security authorities hold different views and make their own catalog. Locally, the number of medicines covered by medical-insurance provincial systems exceeds that of essential medicines stipulated by the national medical insurance system. China's essential drugs list is too large compared with those in countries at comparable economic development levels. There is a huge gap in funds between medical insurance for employees and the new rural cooperative medical care system. In some areas, with the essential drugs system implemented, some patients have turned to secondary and tertiary medical institutions from grassroots medical institutions for medical treatment due to the lack of commonly used drugs or brand-name drugs in community medical institutions. These are also related to the insufficient supply of essential drugs, the unscientific drug selection process, and defective drug tender and procurement policies.

There are many other major problems in China's drug production and circulation, such as unreasonable prices, unscientific approaches for evaluating drug tenders, insufficient innovation capabilities, solidified interest chains in the pharmaceutical industry, and weak supervision.

Current Challenges

People Increasingly Concerned About Health and Putting Forward Higher Requirements for Health Services

The upcoming decade will be a critical period for China to build a moderately prosperous society in an all-round way, a new course for deepening Reform and Opening Up and accelerating the change of the economic growth model, and a new stage for rapid economic and social development. As the people lead a better life and have growing demands for medical and health services, the CPC and the government more emphasize the coordinated development of economy and society and put more efforts in people's livelihood and social security. Medical care and health services have been prioritized in the overall economic and social development. International experience shows that when per capita GDP exceeds US$3000, the upgrading of residents' consumption structure will accelerate, residents' utilization of medical care and health services will increase rapidly, and the requirements for service quality will become higher. With the per capita GDP exceeded US$3000 in 2008, China has entered the period of rapidly upgrading consumption structure, and the people have put forward higher requirements for health services. However, most grassroots health institutions still have problems such as unskillful personnel and lack of equipment. There is a significant gap between their service capabilities and the people's medical service needs.

Medical and Health Expenses Rising Rapidly, Burdensome to Urban and Rural Residents

China's health expenditure has increased in recent years with continuous economic growth. China's health expenditure was CN¥2426.9 billion in 2011, 2.80 times that of CN¥866 billion in 2005. From 2005 to 2011, China's GDP rose from CN¥183.218 billion to CN¥47,156.4 billion, an increase of 2.57 times, which indicates that health expenditure in China grows faster than GDP. Rising prices, widespread application of new medical technologies, increased patient demands, and expansion of the scope and proportion of medical insurance compensations have all had an impact on the growth of health expenditure. In 2014, China's total health expenditure reached CN¥3531.24 billion, an increase of CN¥364.345 billion or 11.5% over the previous year. The proportion of health expenditure in GDP rose to 5.55% in 2014 from 4.73% in 2005.

Meanwhile, China's public health expenditure is relatively insufficient. China's total public health expenditure accounted for 5.55% of GDP in 2014, which was lower than those of Brazil (8.8%) and South Africa (9.2%), close to that of Russia (5.6%), and slightly higher than that of India (4.2%). In Western developed countries, the proportion of total public health expenditure to GDP is usually close to 10%. In general, the Chinese government's investment in this regard cannot fully meet the needs of the people.

Rapid Population Aging and the Elderly's Fast-Growing Medical Care Needs

As medical care and health services advance and people lead a better life, the average life expectancy of human beings has been extended, and aging has become a common social problem in many countries in the world. China has been an aging society. At the end of 2014, there were 138 million people aged 65 and above in China, an increase of 4.55% over 2013, accounting for 10.06% of the country's total population; there were about 19 million people over the age of 80, which is rising rapidly at an annual growth rate of 1 million. It is predicted that China's population aging will see rapid development in the 2020s. By 2025, China's elderly population will reach about 280 million, accounting for about 18% of the total population. It is estimated that by the middle of the twenty-first century, the number of people over the age of 60 in China may reach 400 million, that is, one in every four people; the number of people over the age of 80 will exceed 100 million.

According to statistics, the prevalence of chronic diseases in China is 17%, and that among people over 60 is 2.3 to 3.2 times that of the general population. The per capita medical expenses of the elderly are 2.5 times that of the general population, and the medical expenses paid by people after turning 60 account for more than 80% of their lifetime medical expenses. As economy grows and people lead a better life, the elderly pursue not only access to necessary medical care but also higher quality of life. In particular, families in China are showing an inverted pyramid intergenerational structure of 4:2:1, and the family is weaker for elderly care. Therefore, some empty-nesters, people over 80, and the sick elderly have increasing demands for life care, healthcare, and spiritual comfort. Surveys show that nearly half of the elderly in urban areas live without children, and empty-nesters account for about 40% of the elderly in rural areas. At present, the quantity and quality of elderly care institutions and personnel are far from meeting the needs of society. In developed countries, the average number of nursing home beds per 1000 senior citizens is 50 to 70. China significantly lags behind in this regard and urgently needs to speed up the development. At present, basic care for the elderly with illness, disabilities, and weakness is mainly provided by their family. However, such care is of low quality, and the contradiction between supply and demand is becoming more serious. On the other hand, urban community health institutions focus on patients. Most of the services they provided are diagnosis and treatment of patients. Their work on health enhancement, disease prevention, and vaccinations is still unsatisfactory. There is a big gap between community health care in service content, scale, and quality and the growing demand for health services.

Changes in Diseases and Multiple Health Issues in the New Stage of Economic and Social Development
The new stage of economic and social development brings multiple health challenges, making health undertakings more arduous. Against rapid industrialization and urbanization, China faces accelerating population aging and more complicated health issues. First, the epidemic of major infectious diseases is highly possible. Chronic noncommunicable diseases and mental illnesses pose more threats to people's health. The potential threat of emerging infectious diseases as well as traditional severe infectious diseases should not be neglected. Secondly, changes in the ecology, production, and lifestyle, food and drug safety, occupational injury, drinking water safety, and other issues impact people's health. Frequent natural disasters, accidents, and social security incidents threat medical and health security. The contradiction between the supply and demand of medical care and health services is increasingly prominent, and the service philosophy and models need to be adjusted accordingly.

4.4 Overall Outlook on China's Health Services in 2030

The period from now to 2030 will be an important period for China's social and economic development and the modernization of public services. It is expected that in 2030, China's economic and social development will approach the level of moderately developed countries, and its public services will be mostly modernized. The provision of basic medical and health services will be effectively expanded, and the health service system will be improved. A universal medical insurance system will be established and optimized. All citizens can receive high-quality basic medical security and basic medical care and health services. Medical care and health services will be more accessible, equalized, of higher quality, more efficient, and better received by the public. Individuals' burden of OOP medical expenses will be significantly reduced. The disparities in the allocation of health resources among regions and different groups will be greatly mitigated. All citizens will have access to medical services.

In 2030, Chinese urban and rural residents' health and quality of life will be greatly improved. The main health indicators will reach or be close to the level of major developed countries. Specifically, China's total population will reach 1.45 billion. The average life expectancy is expected to reach 79 to 80 years, an increase of 4 to 5 years compared with 2010. The mortality rate of infants and children under five will be controlled below 6‰ and 8‰, respectively. The incidence of serious, frequent, and disabling birth defects will decrease significantly. The systematic management rate of children will be more than 90%. In 2030, the maternal mortality rate will drop below 12/100,000, and the systematic management rate and percentage of hospitalized delivery of pregnant and lying-in women will be over 99%.

Table 4.1 shows some major indicators that reflect the overall health of the people, with the situation around 2010 as the baseline, the forecast for 2020 as a medium-term reference, and the forecast for 2030 as the target outlook. From the data comparison, we can determine the overall trend of urban and rural residents' health improvement.

In 2020, when a moderately prosperous society in an all-round way is built, the equalization of basic public services will see significant progress. The development gap between regions in China will be narrowed across the board, and so will the gap in basic public services between urban and rural areas. The difference in health status between urban and rural areas and between regions will be further reduced. The public service system will transform from a low-caliber and extensive model to a high-quality and homogeneous model. Urban and rural areas will converge on some key indicators of basic medical care and health services. There will be an optimal mechanism to equalize basic public health services, with richer content. Major diseases and health threats will be effectively controlled. The development of basic medical care and health services will be more balanced, and the distribution of resources will be more rational. The establishment of a high-quality resource-sharing mechanism will be accelerated. The quality of basic medical care and health services in rural areas and former revolutionary base areas, areas inhabited by ethnic groups, remote and border areas and poverty-stricken areas will be significantly improved.

China's medical care and health service system in 2030 will have the following basic characteristics.

Table 4.1 China's main health indicators in 2030 and their longitudinal comparison

Indicator	*Baseline*	*2020*	*2030*
Average life expectancy (year)	74.8 (2010)	77.0	79–80
Mortality rate of children under five years old (‰)	16.4 (2010)	13	8
Urban	7.3 (2010)	5	3
Rural	20.1 (2010)	16	9
Infant mortality rate (‰)	13.1 (2010)	10	6
Urban	5.8 (2010)	4	2.5
Rural	16.1 (2010)	12	8
System management rate of children under three years old (%)	75.0 (2008)	> 80	> 90
Folic acid use rate of pregnant and lying-in women (%)	16.4 (2010)	20	23
Incidence of birth defects (1/10,000)	144 (2008)	70	50
Underweight prevalence of children under five (%)	9(2005)	<5	<2
Maternal mortality rate (1/100,000)	30.0 (2010)	20	12
Urban	29.7 (2010)	10	7
Rural	30.1 (2010)	22	15
Hospitalized delivery rate (%)	97.8 (2010)	> 98	> 99
Urban	99.2 (2010)	99.9	99.9
Rural	96.7 (2010)	99	99.7
System management rate of pregnant and lying-in women (%)	84.0 (2010)	95	> 99
Gynecological examination rate in rural areas (%)	47 (2008)	85	95
Maternal anemia incidence (%)	29	17	10

Source: The Editorial Committee of *Healthy China 2020 Strategic Research Report*, *Healthy China 2020 Strategic Research Report* (Beijing: People's Medical Publishing House, 2012). The data on the "baseline" and "2020" in the other tables in this chapter are mainly from this book

(1) A public health service system with clear division of responsibilities, information sharing, high responsiveness, coordination, and high efficiency will guarantee and promote equal access to basic public health services for urban and rural residents.

(2) A medical service system that meets residents' health needs and features a sound system, a reasonable structure, complementary functions, and close coordination will provide safe, effective, convenient, and affordable basic medical services. A fully competitive market will be formed, and hospitals will be funded by multiple sources. Private medical institutions will develop rapidly and compete healthily with public institutions, bringing better medical services and reducing medical expenses.

(3) A nationally unified, fair, and inclusive national health insurance system will be established and improved, which will cover all urban and rural residents. Medical security will be more advanced, extensive, and intensive. The gap in medical security between urban and rural areas will be further narrowed. The pooled account of medical security will apply at a higher level. Most medical security items will be transferable and sustainable.
(4) China's overall biomedical technology will be close to the world's leading position. Drug lists (including the essential drugs list) will be replaced by a consolidated list under centralized management. The list will be determined and regularly updated by professional drug assessment committees or agencies. A monitoring network and system for drug and medical device safety will be established and improved.
(5) Systems and mechanisms that support the comprehensive, coordinated, and sustainable development of health services will be improved to effectively ensure the standardized operation of the medical care and health system. Nonprofit medical institutions will be well developed. The public welfare support mechanism will be improved. The health industry will be much developed and can basically meet people's multilevel and diversified needs for health services.

From the horizontal comparison with similar indicators of major developed countries (see Table 4.2), we can see the gap between China's main national health indicators and those of major developed countries.

In 2030, the government will better perform its responsibilities in basic medical care and health services. The governments' total health investment will further increase, and the total health expenditure may account for about 9% of GDP. Government health investment will continue to grow faster than recurrent fiscal expenditure, and the proportion of government health input to recurrent fiscal expenditure will increase. After reforms of payment methods and other areas, it is expected that the growth of total health expenditure will be properly controlled in 2030. The burden of medical expenses on individuals will be significantly reduced, and the proportion of personal health expenditure in total health expenditure will be reduced to about 25%, basically at or close to the level of major developed countries (see Tables 4.3 and 4.4).

Table 4.2 International comparison of main health indicators

Indicator		*Country*					
		China	*France*	*U.K.*	*U.S.*	*Japan*	*Australia*
Average life expectancy (2009)	Male	72	78	78	76	80	80
	Female	76	85	82	81	87	84
Maternal mortality rate (1/100,000) (2010)		37	8	12	21	5	7
Mortality rate of children of and under five years old (‰) (2010)		18	4	5	8	3	5
Infant mortality rate (‰) (2010年)		16	3	5	7	2	4

Table 4.3 Health investment indicators

Indicator	*2010*	*2020*	*2030*
Proportion of the total health expenditure in GDP (%)	5.01	6.5–7.0	9
Proportion of the government's health expenditure in GDP (%)	1.19	2.2	4.5
Proportion of the government's health investment in fiscal expenditure (%)	5.3	11.0	15
Proportion of personal expenditure in the total health expenditure (%)	65–75	35	25

Table 4.4 International comparison of health investment indicators (2009)

Indicator		*Country*					
		China	*France*	*U.K.*	*U.S.*	*Japan*	*Australia*
Proportion of the total health expenditure in GDP (%)		5.1	11.9	9.8	17.6	9.5	8.7
Composition of the total health expenditure (%)	Government's health expenditure	52.5	77.9	84.1	47.7	82.3	68
	Personal health expenses	47.5	22.1	15.9	52.3	17.7	32
Per capita health expenses (US$)		191	4840	3440	7960	3754	3945

Around 2030, with economic and social development and the improvement of people's living standards, people will have higher demands for health. The content and standards of basic medical care and health services will be clearer, more extensive, and richer. For example, in terms of basic public health, some programs closely related to public health, such as children's vision care, regular eye and dental examinations for children, cholesterol screening of high-risk groups, and immunization of adults with various high-risk infectious diseases, will be included in basic medical and health services, so will more disease screening and immunization items. In terms of basic medical care, some necessary and appropriate home health services, long-term care services, hospice care, laboratory diagnosis, mental health services and treatment for drug abuse, mental and physical rehabilitation of disabilities or chronic diseases will also be officially listed as basic medical and health services.

4.4.1 Public Health Services

In 2030, with economic and social development and the needs to improve disease prevention and control, there will be more major public health service projects, and public health service capabilities will be enhanced concurrently. A comprehensive public health emergency response system with unified command, reasonable structure, quick response, efficient operation, and strong assurance will be established.

By then, there will be sound professional public health service networks for major disease prevention and control, family planning, and maternal and child healthcare. The monitoring, prevention, and control for seriously life-threatening infectious diseases, chronic diseases, endemic diseases, occupational diseases, and birth defects will be greatly improved. The health supervision system will be upgraded. The standards and risk assessment, monitoring and early warning, and emergency response system for food safety, and the supervision and monitoring system for drinking water hygiene will be completely established. The urban and rural prehospital emergency network will be more mature. The medical rescue, monitoring and early warning, and response for public health emergencies will be greatly strengthened. By then, infectious disease information in China will be timely disclosed, and all public health emergencies will be reported.

In 2030, the public health service model will be transformed. Professional public health institutions will regularly conduct health monitoring and assessments in workplaces, schools, communities, and households. They will also study and develop public health prevention and treatment strategies and guide other medical care and health institutions to provide basic public health services. Urban and rural grassroots medical care and health institutions will engage with families. They will fully know the major health issues in their jurisdiction and among the residents, actively take effective intervention measures, and achieve an organic combination of basic public health services and medical care.

In 2030, information management will play a more vital role in public health. The information system construction in primary medical care and health institutions will be further accelerated. Each urban and rural resident in China will have their own health record, to which primary medical care and health services will have an effective access. Information technology will further optimize the management system and procedures, promote public health service efficiency and management, and improve the service quality and management effects.

In 2030, immunization will be more accessible. The greatly strengthened IT-based management will make it more convenient for the floating population to get vaccinated. Up to about 99% of children recommended to get vaccinated (including the floating population) will receive vaccination under the national immunization program. The spread of major and sudden infectious diseases such as HIV/AIDS will be effectively curbed. The infection of viral hepatitis, tuberculosis, and insect- and animal-borne diseases will be significantly reduced. By then, iodine deficiency disorders will be generally eliminated in nearly all Chinese counties (cities and districts) . In addition, the Kashin-Beck disease, the Keshan disease, endemic fluorosis caused by coal-burning pollution, and arsenic poisoning caused by coal-burning pollution will be basically eradicated (see Table 4.5).

In 2030, health education will be fully integrated into basic public health services. Internet and mobile communications will be widely used in public health education. Personalized health guides will be popularized. Appropriate health education programs will be developed for key groups to increase their engagement and compliance.

China's population aging will develop rapidly around 2030. Grassroots medical care and health institutions will focus on people over 65, providing them with physical examinations, disease screening, health consultation and guidance, chronic disease management, and rehabilitation. By then, chronic disease screening and tracking will be integrated with outpatient services by grassroots medical care and health institutions. The institutions will increase the screening and detection of patients with

Table 4.5 Indicators for infectious disease control

Indicator	*Baseline*	*2020*	*2030*
Vaccine-preventable diseases in the Expanded Program on Immunization	> 14	20	> 25
Increasing the coverage of the child immunization plan (coverage of individual vaccines)	90%	95%	99%
Reducing the HBsAg carrier rate in the population	> 7.18%	< 6%	< 4%
Reducing the incidence of measles	3/100,000	0.1/100,000	0.01/100,000
Increasing the coverage of AIDS anti-infective treatment	70%–80%	100%	100%
Curbing and reversing the spread of AIDS	700,000	< 1,500,000	< 1,800,000
Reducing the prevalence of smear-positive TB	122/100,000	50/100,000	30/100,000
Reducing schistosome infections among residents (in endemic areas)	< 5%	No new cases	No new cases
Improving the standardized diagnosis and treatment rate of malaria patients	80%	100%	100%
Reducing the incidence of malaria	< 1/10,000	No domestic infection cases	No domestic infection cases
Dog deworming rate	30%–40%	> 90%	> 97%
Standardized diagnosis and treatment rate of hydatid patients	40%	> 90%	> 97%
Reducing the seroprevalence of the hydatid disease among children	> 15%	< 5%	< 2%
Standardized deworming rate of soil-transmitted nematodes	30%	> 90%	> 96%
Reducing the infection rate of soil-transmitted nematode	20%	5%	2%
Reducing the prevalence of iodine deficiency disorders	Prevalent to varying degrees	Eradicated	Eradicated

hypertension and diabetes, accurately grasp patents' conditions, and enhance their awareness. At the same time, the institutions will promptly include the diagnosed patients in health management, provide standardized services according to the patient's condition, and strengthen better guidance on lifestyle and medication. In 2030, the spread of chronic non-communicable diseases will be effectively controlled. Prevention will be combined with treatment and whole-course management for chronic diseases, significantly reducing the harm caused by chronic diseases and the mortality rate of chronic patients (see Table 4.6). Improvement of rural health will also be a priority. Table 4.7 shows the international comparison of rural health indicators.

Table 4.6 Indicators for the control of health risk factors

Indicator	*Baseline*		*2020*		*2030*	
Reducing smoking prevalence in the population (over 15 years old)	*Male: 53%*	*Female: 4%*	*Male: 40%*	*Female: 4%*	*Male: 30%*	*Female: 3%*
Reducing overweight among children	9% (2005)		6%		5%	
Reducing overweight among adults	30.6% (2010)		25%		20%	
Improving the rate of indoor air quality meeting the standard	Setting the baseline		80%		90%	
Increasing the coverage of safe drinking water in rural areas	86% (2008)		98%		100%	
Promoting furnace and stove improvement in areas of diseases caused by coal-burning pollution	Setting the baseline		100%		100%	
Increasing the coverage of sanitary toilets in rural areas	43% (2008)		90%		100%	
Proportion of residents with personal health files	20%		90%		98%	
Improving the management and control rate of patients with hypertension	< 10%		50%		80%	
Improving the management and control rate of patients with diabetes	< 10%		40%		70%	
Improve the management rate of severe mental illness	11%		80%		95%	
Reducing stroke mortality	137/100,000 (2005)		Reduced by 20%		77/100,000	
Reducing cardiovascular mortality	90/100,000 (2005)		Reduced by 20%		50/100,000	
Reducing esophageal and gastric cancer mortality	40/100,000 (2005)		Reduced by 30%		20/100,000	

Table 4.7 International comparison of rural health indicators

Indicator	*Country*					
	China	*France*	*U.K.*	*U.S.*	*Japan*	*Australia*
Population (10,000)	1,348,932	62,787	62,036	310,384	126,536	22,268
Proportion of urban population (%)	47	85	80	82	67	89
Coverage of safe drinking water in rural areas (%)	91	100	100	99	100	100
Coverage of sanitary toilets in rural areas (%)	64	100	100	100	100	100

4.4.2 *Medical Care and Health Services*

In 2030, China's total health resources will increase substantially. Their structure will be more reasonable and the service system will be sounder. Health services will be more equal and accessible. The number of beds in medical care and health institutions per 1000 permanent residents will increase to about 7.7, the number of licensed (assistant) physicians per 1000 permanent population will rise to about 3, the number of registered nurses per 1000 population will be close to 6, and the number of public health personnel per 1000 population will reach 1.1. The number of personnel will meet Chinese people's health service needs. The distribution of medical and health personnel between urban and rural areas and between regions will be more reasonable. Different medical personnel groups will develop in a coordinated manner.

In 2030, an urban medical care and health service system based on community health services will be fully established and improved, which will coordinate community health service institutions, clinics, hospitals, and preventive health institutions, and the three-level medical and health service networks in rural areas. In cities, 98% of residents can reach the nearest medical institution within 10 minutes (including by car). In rural areas, this ratio will rise to about 88% (see Table 4.8). An unblocked, coordinative, and standardized working mechanism for division of responsibilities and cooperation will be established among professional public health institutions, urban and rural grassroots medical and health institutions, and hospitals, to ensure information and resource sharing and combining prevention with treatment.

Table 4.8 Equity and accessibility of health services

Indicator	*Baseline*	*2020*	*2030*
Strengthening grassroots network construction and expanding coverage of village clinics	88%	95%	100%
Rate of residents with reach to the nearest medical institutions within 10 minutes (including by car)	Urban: 80% Rural: 65%	Urban: 90% Rural: 75%	Urban: 98% Rural: 88%
Coverage of essential drugs	30%	100%	100%
Number of physicians and physician assistants per 1000 population	1.58	2.5	3.0
Number of registered nurses per 1000 population	1.25	3.14	6
Number of public health personnel per 1000 population	0.61	0.83	1.1
Number of general practitioners per 10,000 population	1.07	2	3.3
Number of hospital beds per 1000 population	3.05	6	7.7
Rate of unhospitalized patients	25%	8%	3%
Average length of stay in hospital (day)	8.6	7	6

Moreover, the overall capacity of medical care and health professionals and the skills of grassroots medical personnel will be significantly improved. The workforce with general practitioners (GPs) as the main body will mature and be stronger. GPs will establish a stable service relationship with urban and rural residents. Basically, every 10,000 residents in urban and rural areas will be served by two to three qualified GPs. GPs' service will be improved in an all-round way, basically meeting people's needs for basic medical care and health services. China's medical care as a whole will be close to that in major developed countries, and some specialties will advance globally. As the regulation system continues to be improved and the gap between urban and rural areas continues to narrow, everyone can receive high-quality basic medical care and health services, and convenient, affordable, safe, and effective access to medical care.

Table 4.9 shows that there is still a gap between China and developed countries in terms of equity and accessibility of health services.

By 2030, major breakthroughs will be made in the public hospital reform. The administrative rankings of public medical institutions at all levels will be basically removed. Various forms of corporate governance structures and a professional manager system will be established. The management system, corporate governance mechanism, and compensation mechanism for hospitals and the classified management system for medical institutions will be sounder and more mature. A comprehensive

Table 4.9 International comparison of equity and accessibility of health services

Indicator	*Country*					
	China	*France*	*U.K.*	*U.S.*	*Japan*	*Australia*
Number of physicians per 1000 population 2005–2011	1.42	3.45	2.74	2.42	2.14	2.99
Number of hospital beds per 1000 population 2005–2011	4.20	6.90	3.30	3.00	13.70	3.80
Number of nurses per 1000 population 2005–2011	1.38	0.31	10.13	9.82	4.14	9.59

hospital evaluation system for the public good will be established and used to direct the government's supervision and regulation over medical institutions. Special service beds and facilities in public hospitals will no longer exist. The internal income distribution system of public hospitals based on job workload, service quality, and patient satisfaction will be sounder.

Moreover, the personnel and salary system will be reformed to adapt to the characteristics of the medical and health industry. Medical workers at public medical institutions at all levels will no longer be managed as public institution staff. Physicians can practice at multiple sites, working as full-time or part-time hospital employers or setting up their own clinics. The comprehensive medical training system will be further improved, transforming from institution-based training to open training. The reform of the salary system and total salary system of public medical institutions will reasonably increase the income of medical workers. A salary system that effectively reflects the value of medical and health workers will be established. On the other hand, a social assessment and information disclosure system for medical service institutions and practicing physicians will be established to strengthen the medical supervision mechanism of information disclosure and social engagement.

In many parts of the country, a community- or GP-based primary care systems will be established. Family doctors will better serve as the "health guardian." The hierarchical diagnosis and treatment and two-way referral system will be more mature in economically developed regions. GPs will have the right to participate in the medical process after the patient is referred to ensure the continuity of treatment and medication. They will also have the right to be involved and intervene in the treatment of the patient by referring medical institutions, as well as the rights concerning the patent transferred back to the community after diagnosis and treatment. Medical groups similar to regional medical unions or aiming to integrate medical and health services will be promoted in some regions to build a convenient green referral channel for patients with such needs.

Community health services will be the focus of medical model transformation. The general practice team will include psychotherapists, medical social workers, and other professionals, becoming a medical care and health service team comprising general practitioners, community nurses, psychotherapists, and social workers. The resources of community health service institutions, community life service centers, community cultural centers, and other grassroots service institutions will be effectively integrated into an important support for community health services.

China's medical service market in 2030 will be a fully competitive market, in which health resources will be scientifically and rationally allocated and guaranteed by an effective mechanism. Private medical institutions and public medical institutions will be in healthy competition. The number of beds and services of social medical institutions will be close to 40% of the total medical services. Nonprofit medical institutions will be highly developed, and there will be a well-established mechanism to ensure that they are for the public good.

By then, medical and health information technology will be more advanced. Telemedicine and mobile health will be greatly developed, and will promote fundamental changes in the medical service pattern and medical treatment practices. Digital platforms and transparent systems will allow patients to learn about the educational background, experience, charges, reputation, and reviews of a doctor. Patients can openly rate a doctor immediately after seeing him or her on digital platforms. Doctors who truly deserve the respect of their peers and patients will be recognized and appreciated. Doctors will be paid more reasonably based on their medical skills and services.

In 2030, patients will see a doctor via telemedicine in most cases instead of visiting a hospital, and their medicines will be timely delivered through an advanced logistics system. Mobile medical care can maximize the use of medical and health resources on the mobile Internet and extend the access to medical services to more people. Communication networks will also be used to enhance medical care and health services, including basic care, public health research, emergency care, chronic disease management, and self-service medical care, greatly improving the operational efficiency of the medical care and health system. People's health data will be permanently recorded in the database of a national health information system. Patients can be diagnosed at home. Based on health data, doctors can remind patients of changing their behaviors and suggest them on health enhancement and physical activities.

4.4.3 *Medical Security and Assistance*

In 2030, a nationwide-unified, fair, and inclusive health insurance system will be established and improved, which will cover all urban and rural residents. Everyone will be covered by basic medical insurance. There will be no difference between urban and rural areas in the pooled account and management of basic medical security, which means that all citizens can enjoy medical security fairly and fully.

The medical insurance system will be transformed into a health insurance system. Regional institutional arrangements will develop into nationally unified institutional arrangements. Under a clear responsibility-sharing mechanism between the central and local governments, regional health insurance systems for citizens will remain open. Capital, workers, and residents will be allowed to move freely between regions to motivate local governments to improve their health insurance management and services. Service purchase and delivery systems in the health security system will be integrated to encourage medical institutions turn their focus from treatment to health security that pays equal attention to prevention, treatment, and rehabilitation. A science-based and standardized fee payment and service purchase mechanism will be formed. In medical services, sufficient market competition will be promoted to improve service quality.

Around 2030, the medical insurance funds may no longer set payment caps. The proportion of personal health expenditure in the total health expenditure will drop to about 25%, which means that the personal burden of patients will be further relieved. The urban and rural medical assistance system will be further improved to cover all poverty-stricken households. There may be no more cases of falling into poverty because of catastrophic medical expenses (see Table 4.10). At the same time, commercial medical insurance will be more mature and become an important supplement to basic medical insurance.

Furthermore, as China's population ages, long-term care will see increasing demand and will become a health service as important as

Table 4.10 Mitigating disease-related economic risks

Indicator	*Baseline*	*2020*	*2030*
Expanding the coverage of basic medical insurance	Rural: 90%, urban: 72%	100%	100%
Increasing the benefit rate of basic medical insurance: proportions of personal expenditure	Rural: 75%, urban: 65%	34%	25%
Improving the urban and rural medical assistance system: coverage for poverty-stricken households	Setting the baseline	100%	100%
Eliminating cases of falling into poverty because of medical care (catastrophic medical care and health expenses)	10%	2%	< 0.5%

general diagnosis and treatment, disease prevention, and health management. Around 2030, long-term care will likely become one of the basic medical and health services in China. A basic care security system will be established and improved in China. The government will build a safety net with care assistance. A care insurance system with long-term basic care insurance as the core and care insurance and commercial care insurance as supplements will be formed. Home care and community care will be supported by the government and society. That way, a professional, socialized, and multilevel care service system will be formed.

4.4.4 *Drug Supply Assurance*

In 2030, after rapid development, China's pharmaceutical industry will perform much better in independent innovation. China's biomedical technology as a whole will be close to the world's leading ranks. The biomedical industry has become one of the pillars for the national economy.

By then, a drug assessment committee or professional drug assessment agency comprised of clinicians, health economists, and payer representatives will determine a drug list and regularly update it, and set the principles, standards, procedures, and methods for drug selection. The agency will fully consider the effectiveness, safety, quality stability, cost-benefit comparison, social value, and production supply assurance of the drugs based on national clinical diagnosis and treatment guidelines. The selection will be open and transparent, and the opinions of grassroots medical workers will be fully listened to. The selected drugs will meet basic medical needs.

At the same time, the drug pricing mechanism will be more scientific and reasonable. A drug pricing committee, which is composed of government officials responsible for drug pricing, the expert advisory committee for drug pricing (composed of experts in drug research and development, drug production and quality control, and drug price research, clinical experts, and pharmaco-economists), pharmaceutical companies, pharmaceutical companies, medical insurance departments, hospitals, and consumer representatives, will discuss and determine the initial prices of drugs. A drug cost monitoring system and a science-based and reasonable joint adjustment mechanism for costs and prices should be established and improved. A drug cost monitoring network will be established, which will collect information on the cost and fluctuations in supply and demand of active pharmaceutical ingredients, technical difficulties, and changes in pharmaceutical production and grasp the legitimate expenditures in the circulation of pharmaceuticals to provide a basis for the final calculation of the average social cost of drugs. It will also account for the costs of bringing innovative medicines to market. In terms of pricing models, pharmaco-economics will be fully utilized, and drug prices will be set through an integrated approach.

By then, China's pharmaceutical industry will be more concentrated. China's position in the global industry will undergo major changes, from a global manufacturer to the most innovative and important R&D center and the most promising market. Pharmaceutical production will be highly intelligent. The application of human-computer intelligent interaction, industrial robots, big data, and cloud computing will be very common.

In 2030, a monitoring network and system for drug and medical equipment safety will be established and improved in China. All drugs and food will be subject to safety testing. More than 98% of food additives in use can be justified. China's drug quality pass rate and drug safety will be comprehensively improved (see Table 4.11). The public's satisfaction with food and drug safety will be greatly improved.

4.5 Paths and Policy Choice

4.5.1 *Path to the Equalization of Basic Medical Care and Health Services*

The equalization of basic medical and health services refers to science-based coordinating and developing public health and basic medical

Table 4.11 Food and drug safety indicators

Indicator	*Baseline*	*2020*	*2030*
Drug and food safety testing coverage	Setting the baseline	100%	100%
Chemical contaminant exposure rate of food	60%	40%	20%
Rational use of food additives	Setting the baseline	90%	98%
Hazard coverage of the risk assessment database	Setting the baseline	90%	98%
Incidence of food poisoning	Setting the baseline	Reduced significantly	Reduced significantly
Antibiotics/Intravenous therapy	60%	40%	20%
Incidence of adverse drug reactions	421/1,000,000 population	Reduced significantly	Reduced significantly
Drug safety incidents	Setting the baseline	Reduced significantly	Reduced significantly

services within the government's financial capacity. On the premises of ensuring equal opportunities and access to services for residents, providing services as needed, and respecting residents' right of free choice, basic medical care and health services should be provided to subjects with equal needs in a basically equal manner under the principles of equity and impartiality. The content of services should be determined according to the comprehensive social and economic development, and the capacity and demands of the supply and demand sides of medical and health services. Services should be aimed at equality under certain constraints and changed with economic and social development and residents' health and income. Based on various research conclusions, in order to achieve equalization of basic medical care and health services in 2030, China should:

Clearly establish the dominant role of the government and reasonably define government administrative power and expenditure responsibilities. Experience of China and other countries shows that the market, however powerful, cannot guarantee balanced public resource distribution. Therefore, we should first clearly establish the dominant role of the government. We should make strategic adjustments to public health resources between urban and rural areas and between regions to promote rational public health resource distribution. Particularly, we should guide and promote existing urban health resources to radiate and transfer to surrounding rural areas to realize urban-rural resource sharing. We should reasonably define the central and local governments' administrative power and expenditure responsibilities for basic public services. For the basic medical and health services most concerned and urgently needed by the people and with obvious external spillovers, we should completely change the system in which local governments assume the main responsibility for public health expenditures. We should strengthen transfer payments and invest more effective fiscal funds in former revolutionary base areas, areas inhabited by ethnic-minority groups, remote and border areas, poverty-stricken areas, and disadvantaged groups. The central and provincial governments should pay more to relieve the expenditure burden on the governments below the provincial level, especially county and township governments, so that they have the basic financial capacity to provide basic medical and health services.

Underline the concept of "public funds for people's livelihood," steer the use of funds, and appropriately increase medical and health expenditures. Due to differences in economic development and fiscal revenue and expenditure, different parts of China vary significantly in the scale of medical and health expenditures and the provision capacity and quality of basic medical and health services. Therefore, the central government should increase spending in medical care and health, and local governments at all levels should increase public funds for medical care and health to provide financial support for developing local medical care and health services and equalizing basic medical care and health services.

Optimize the use of medical care and health expenditure and shift the focus of China's medical care and health funds from treatment to prevention. Since the beginning of Reform and Opening Up, some hospitals have started to chase profits in China's medical system reform. They have invested heavily in the highly profiting clinical medical treatment but ignored early prevention and diagnosis, which can effectively reduce the morbidity and mortality of diseases and improve residents' health. In reality, 80% of China's deaths are due to chronic diseases, which requires China's medical and health services to shift from treatment-oriented to prevention-oriented services. The focus of medical care and health expenditure should change accordingly.

Effectively build a strong grassroots health workforce and improve the basic public health service capacity. In recent years, grassroots health services have attracted much attention. The central government has introduced policies to strengthen the grassroots health service system.

Nonetheless, there is still no substantial mandatory policy for the grassroots health workforce. The shortage of professionals in the grassroots health service system remains. A large number of public health services cannot be put in place. Therefore, we should focus on strengthening the training of grassroots professional health personnel and building a grassroots medical care and health workforce with a focus on general practitioners. We should carry out job transfer training for general practitioners, nurses, and traditional Chinese medicine personnel. Besides scientifically verifying the staffing and increasing the number of public health personnel, we should improve the professional capacity of public health personnel and train more public health technical and management personnel to improve the capacity of grassroots public health services and management. We should further raise the remuneration for grassroots professional health personnel to attract and stabilize the grassroots health service workforce and ensure that basic medical and health services are provided in the quality and quantity as required.

Continue to increase investment in infrastructure construction to enhance service capacity. We should focus on improving service networks, enhancing service capacity and optimizing systems and mechanisms, to further improve urban and rural health infrastructure. We should highlight and strengthen the functions of basic medical and health services. We should focus on building township health centers and other grassroots institutions, such as community health service clinics and village clinics, and strengthen equipment allocation. That way, we will comprehensively promote the standardized construction grassroots public health service institutions.

Strengthen the grassroots health service system and transform grassroots medical and health institutions' operation mechanism. We should strengthen the three-level medical and health service networks in rural areas, which are led by county hospitals and based on township health centers and village clinics. We should also improve the new medical care and health service system in urban areas based on community health services. We should increase financial investment. The government should pay for the basic construction, equipment, personnel, and public health service operations of public grassroots medical and health institutions, subsidizing through specific projects and service purchases at the prescribed amount. We should speed up the roll-out of hierarchical diagnosis and treatment and the two-way referral system, divide responsibilities between grassroots medical and health institutions and urban hospitals, and promote multisite practice and free movement of medical personnel. We should clearly define grassroots medical and health institutions' public service function so that they can truly operate for the public good. We should establish a sound performance appraisal mechanism focusing on service quality and social benefits.

Improve medical security and the applicable level of pooled accounts. We should effectively increase medical insurance systems' compensation and payment caps and expand the pooled account for outpatient bills. We should further improve the new rural cooperative medical care system's financing and narrow the gap between urban and rural areas. We should increase the compensation for major diseases and reduce the economic burden on farmers to see a doctor. We should properly connect different medical insurance systems, integrate operation resources, and improve pooled accounts' applicability. We should accelerate cross-regional medical insurance relation transfer and continuation and medical expense settlement, as well as the integration of urban and rural medical insurances.

Accelerate a pattern with multiple hospital investors, and encourage, support, and guide private capital into medical care and health services. In regional health planning and medical institution setup planning, we should leave reasonable space for private medical institutions. Nonpublic hospitals should enjoy fair treatment in access, practice, supervision, medical insurance designation, scientific research project approval, and employees' title evaluation and further education. We revise and improve preferential tax policies and guide nonpublic hospitals to operate according to law.

4.5.2 Reform and Innovation of the Medical Care and Health Service System

We should establish a national health administrative system and integrate medical care, health, and social welfare affairs into a "general health" administrative system that coordinates medical security and services. We should improve the consultation and accountability mechanism for health decision-making and organize experts for necessity and feasibility demonstration before major decision-making. For decisions that involve many stakeholders and are closely related to the people's interests, we should publicize them and implement a hearing system. We should improve the accountability mechanism for decision-making, establish an accountability mechanism for industry associations, and incorporate public health indicators into the key assessment indicator system for governments at all levels.

We should optimize medical resource allocation. We should insist on the dominant role of nonprofit medical institutions and the complementary role of for-profit medical institutions, and public medical institutions leading and developing side by side with nonpublic medical institutions. We should formulate regional health plans and medical institution setup plans based on people's needs, adjust the distribution and structure of medical resources according to population distribution and flow trends, and reasonably determine public hospitals' function, quantity, scale, structure, and layout. We should curb public hospitals' blind expansion. When the number of beds in medical and health institutions per 1000 permanent residents reaches four, public hospitals will not, in principle, be expanded. We should effectively meet the medical resource needs in remote areas, new areas, suburban areas, and satellite cities. We should strengthen the medical service capacity in pediatrics, obstetrics and gynecology, mental health, oncology, infectious diseases, elderly care, rehabilitation medicine, and traditional Chinese medicine. New medical and health resources should focus on weak links in rural and urban communities to ensure access to basic medical services. We should vigorously develop continuous medical institutions such as rehabilitation hospitals and nursing homes (stations) and improve their rehabilitation medical services and nursing capacity. We should strengthen the medical service system and improve the medical services for women and children. We should boost the construction of provincial specialized maternal and child health hospitals and the establishment of pediatric departments (specialized hospitals) in provincial and prefecture hospitals, as well as in the establishment of gynecology and pediatric departments in county hospitals. We should strictly control the allocation of large medical equipment and encourage joint construction and sharing to improve the efficiency of medical and health resources. We should guide patients to seek medical treatment reasonably and ensure that they have access to high-quality medical services nearby.

We should upgrade the compensation mechanism for public hospitals and implement a government investment policy. Government-funded medical institutions will no longer be subsidized on a per capita basis or on a loss basis. A compensation mechanism based on performance appraisal results will be established to encourage them to provide better services. A classified subsidy system will be in place to improve the assurance and support policies for medical institutions for infectious diseases, psychiatry, pediatrics, and traditional Chinese medicine. The focus of government investment will shift from securing the supply side to subsidizing the demand side. We should increase the investment in medical insurance, rationally adjust the direct investment in medical institutions, and improve the government purchase of service mechanism accordingly. We should advance the medical service price reform to eradicate the practice of charging more for medicines to make up for low prices for medical services. That way, we can curb overmedication, over-examination, and overuse of high-value consumables through the benefit mechanism, and reasonably relieve the burden of patients seeking medical care. We should adjust the prices of medical services under the principle of being proactive and prudent, and prioritizing mature approaches. We should explore new medical price management methods, such as disease-specific pricing, package pricing, and negotiated pricing under medical insurance, and establish a pricing mechanism conducive to standardized medical behaviors and higher service quality.

We should remove the administrative rankings of public hospitals. We should strengthen health administrative departments' industry administration functions, including planning, admission, and supervision. The independent operation and management rights of public hospitals should be exercised, and administration and operation should be split. We should improve the governance mechanism of public hospitals and explore various forms of corporate governance such as councils. We should establish a professional manager system and implement an employment system for professional presidents and a tenure target accountability system.

We should reform the personnel management system, turning the medical personnel from "public sector workers" to "employees." Eligible medical workers at government-funded medical institutions can work part-time at private medical institutions or set up private clinics. We should promote and improve the standardized training system for residents. We should promote the reform of the salary system and total salary system for public medical institutions, and establish a salary system that reflects the value of medical and health workers. At the same time, we should implement an annual salary system for hospital presidents based on the public welfare performance appraisal results. That way, we will reasonably increase the income of medical workers. We should improve the medical practice insurance system and implement compulsory medical liability insurance. In addition, we should establish a social evaluation and information disclosure system for medical service institutions and practicing physicians, and enhance the supervision mechanism for information disclosure and social engagement.

We should define the functional positioning of public hospitals and grassroots medical and health institutions, and set a hierarchical diagnosis and treatment system according to the service functional positioning, that is, a pyramid medical treatment model. The top or spire of the pyramid is national key disciplines, key clinical specialties, and clinical medical research centers. Their main tasks are medical scientific research and advanced personnel training, and their functional positioning is strengthening scientific research and personnel training. The middle part of the pyramid is made up of regional medical centers, which are for the treatment of emergencies and severe diseases in the region. The base of the pyramid is the networked and widely distributed community medical and health institutions and private clinics, which combine prevention with treatment and serve as the "health guardian" for residents. They mainly serve for the treatment of common and chronic diseases, rehabilitation of severe illnesses, disease prevention and control, and health promotion. We should strengthen the support and guidance of public hospitals to grassroots medical and health institutions. We should strengthen the division of responsivities and cooperation among medical institutions and form a diagnosis and treatment model of primary care at grassroots institutions, hierarchical medical care, coordinating different levels, and two-way referral.

We should seek to form regional medical groups or medical unions for convenient referral channels for patients with such needs. We should advance the reform of payment methods and speed up the exploration of the unified prepayment of medical insurance for medical institutions in one union. We should implement period-based payment for hospitalized patients, and capitation for community health services. We should widen the gap in the reimbursement ratio for medical treatment between medical institutions at different levels, and the gap in the medical reimbursement ratio between contracted and noncontracted groups.

We should organically integrate health services and social services. We should establish a community service administration department in which multiple government departments participate and collaborate, and regard the grassroots and the community as significant levels for health issues. We should seek opportunities and methods to integrate social and public services provided by health and civil affairs departments and China Disabled Persons Federation, so as to effectively connect and integrate health services with various social public services and provide services that adapt to the characteristics of the aging population and the needs of local residents, and are more user-friendly, convenient, and effective.

We should drive the reform of medical models in the community. We should seek the organic integration of community health services based on the general practitioner system and public services such as social services, community elderly care, and elderly assistance in the community, with community health services as the core. We should deeply integrate community basic medical big data with mobile medical and telemedicine platforms to achieve the IT-based integration of disease prevention, treatment, rehabilitation, and health management.

We should control the unreasonable increase of medical expenses. We should tighten the supervision over medical expenses and incorporate the targets of the control and management of the growth rate of average and total expenses, length of stay in hospital, and drug sales as a share of total revenue for hospitals into the target management accountability system of public hospitals and take them as important indicators for performance appraisal. We should promptly investigate and deal with unreasonable drug use, material use, and repeated examinations for extra economic interests. We should strengthen the key monitoring of the diagnosis and treatment of diseases with rapidly increasing costs and control the provision of nonessential medical services by public hospitals.

We should tighten the supervision over medical services. We should establish a sound medical service supervision system, revise medical service supervision laws and regulations, and strengthen the monitoring and supervision over medical service practices, quality and safety, and institutional operation. We should upgrade the complaint management system, the medical dispute mediation mechanism, and the medical liability risk-sharing mechanism. We should upgrade the hospital grade approval and evaluation system and strengthen regular quality control evaluation. We should comprehensively promote the regular assessment of physicians to regulate medical practice. We should expand public engagement and establish a long-term mechanism for social supervision and evaluation with government agencies and authorized professional organizations as the primary supervisor. Regulators should include not only government representatives, payer representatives, medical experts, but also public representatives and legal experts. With a group of broadly representative supervisors, transparency and accountability of hospitals, including public hospitals, will improve.

The supervision approach to improve medical care shall feature two major aspects. The first is to develop specifications and guidelines for diagnosis and treatment. Policymakers and professional groups should work together to develop medication specifications and diagnosis and treatment guidelines following evidence-based medicine to help funders and regulators oversee providers. The second is to publicize the results of medical institutions' clinical services. For example, information about medical service providers' performance should be disclosed and released to allow patients and their proxies to choose high-quality services and to encourage providers to improve service quality. Information management systems should also be fully utilized to oversee and evaluate health service institutions' practice and the health system's performance as a whole. For example, an information management system should be established in the medical insurance designated hospital system to record hospital services and reduce the difference in diagnosis and treatment. Or a pay-by-disease information management system should be established to improve management and the comparability of service quality among hospitals.

We should vigorously develop nonpublic medical institutions. We should leave enough space for nonpublic medical institutions in regional health planning and medical institution setting planning. When it is necessary to adjust and add medical and health resources, we should prioritize private capital as allowed by the access standards. We should introduce more private capital to set up medical institutions and encourage capable enterprises, charities, foundations, commercial insurance institutions, and overseas investors to set up medical institutions, as well as qualified individuals (including those from Hong Kong, Macao, and Taiwan) to set up private clinics legally. Cities with abundant public hospital resources can guide private capital into restructuring and reorganizing some public hospitals, including hospitals run by state-owned enterprises, in various ways. The cities may reorganize some public medical institutions into nonpublic medical institutions actively while steadily and moderately reduce the proportion of public hospitals. The layout of public hospitals should be improved to form a pattern with multiple hospital investors. We should revise policies and measures to incentivize nonpublic medical institutions and further improve the practice environment. We should implement policies on price, hospital designation in medical insurance, land, key discipline construction, professional title evaluation, and large equipment allocation, and give priority support to establishing private-funded nonprofit medical institutions. We should implement preferential tax policies for nonprofit medical institutions and revise tax policies for for-profit medical institutions.

We should encourage and guide private capital into the construction, operation, and management of medical and health service facilities. We should open up the access of basic public services fairly and encourage social capital to set up elderly (day) care service institutions and rehabilitation and nursing service institutions for people with disabilities. In basic public service provision, we should practice government purchase, franchising, contract entrustment, and service outsourcing. Private basic public service providers shall be treated equally with public institutions in terms of establishment conditions, qualification recognition, taxation policies, government purchase of service, and evaluation of personnel's professional qualifications and titles. We should seek financial funds for nonpublic basic public service institutions, increase citizens' flexible choice of services, and promote fair competition among basic public service institutions by providing direct financial subsidies to the demand side. In addition, it is recommended to develop measures for nonprofit institution management to constrain their practice, including the allocation and use of equipment, facilities, drugs, and services, the proportion of low-income patients and the medical services they receive, and the support for public welfare undertakings.

4.5.3 Improving the Public Health Service System

We should establish a coordinated and unified medical and healthcare management system, integrate health management functions, strengthen overall coordination, and improve administrative efficiency. We should enhance government leadership in the provision of public health and basic medical services. We should improve the government health investment mechanism. Government health investment should grow faster than recurrent fiscal expenditure and occupy a higher proportion in recurrent fiscal expenditure. We should reasonably divide the health expenditure responsibilities among the central and local governments at all levels.

We should standardize and improve public health service management. We should more widely apply the National Basic Public Health Service Standard and better manage major public health service projects. We should integrate existing major public health service projects and implement them as a whole. We should establish an expert demonstration mechanism for major public health service projects, highlight the keys of different phases, stress the evaluation of project implementation effects, and practice dynamic management. We should speed up the construction of community health service institutions and township health centers (village clinics), and manage urban and rural grassroots preventive health care based on networks. A working mechanism with clear division of responsibilities, complementary functions, information and resource sharing, and coordination and interaction should be established among public health service providers for the combination of prevention with treatment.

We should improve public health services. We should optimize public health staffing according to the standards, train more technical and management personnel for public health services, provide public health knowledge and skill training for general practitioners, community nurses, and rural health personnel, and stabilize the public health service workforce at grassroots medical and health institutions. We should fully mobilize the workers by reforming personnel and performance management and the income distribution system. We should strengthen the monitoring, early warning, and emergency response of major diseases and public health emergencies. We should establish a sound public health service information system mainly for disease control, maternal and child healthcare, health supervision, community health, and rural health. We should enhance information standardization and the public health service information platform and create standardized and unified residents' health e-records to improve public health service efficiency and management.

We should strengthen the major disease prevention and control system. We should monitor key diseases, strengthen the construction and management of the infectious disease direct reporting network, and improve the disease monitoring system and information management system. We should strengthen the laboratory testing network system for disease prevention and control and establish a quality management system for infectious disease laboratories. We should continue to prevent and control major infectious diseases, parasitic diseases, and endemic diseases. We should improve the regular vaccination rate of immunization programs and the quality of vaccination management of floating population, and boost the vaccination of key groups.

We should speed up improving the public health emergency response system. We should improve the information reporting, risk assessment, and monitoring and early warning systems, and also improve the prevention and control of public health emergencies. We should make comprehensive health response to major public emergencies and active emergency medical response to major disasters and accidents, focusing on on-site medical and health rescue in disasters, health emergency response to poisoning and nuclear radiation incidents, and psychological assistance for emergencies. We should provide strong health support for major events. We should strengthen national and provincial emergency medical rescue and laboratory emergency testing, and particularly support health emergency response teams in Central and West China. We should strengthen the prehospital emergency system and, particularly, improve the emergency medical service capacity in rural areas.

We should fully implement the strategy of comprehensively chronic disease prevention and control. We should strengthen the detection and preventive intervention of chronic diseases among high-risk groups with comprehensive grassroots prevention and control of hypertension, diabetes, and other chronic diseases. We should vigorously carry out the Healthy Lifestyle for All campaign, set up demonstration zones for comprehensive chronic disease prevention and control, and provide health management, lifestyle guidance, and intervention for high-risk groups. We should strengthen the screening and prevention of stroke, coronary heart disease, and other cardiovascular and cerebrovascular diseases. We should carry out key cancer screening and early diagnosis and treatment in areas with high cancer incidence. We should establish a chronic disease prevention and control system for both urban and rural areas.

We should establish a sound and fully functional management and treatment network for severe mental illness for both urban and rural areas. We should establish a case reporting system for severe mental illness, strengthen management and treatment, and ensure antipsychotic treatment and emergency assistance for impoverished patients with severe mental illness. We should improve social psychological support and the mental health service system, and strengthen regulated and standardized management.

We should extend health education. We should fully leverage the health education system and health education centers and design health education activities on key diseases, groups, and places, as well as major public health issues. We should continue to carry out the Health Literacy Promotion for All campaign, spread basic health knowledge, increase tobacco control communication, and advocate a healthy lifestyle.

We should provide high-quality public health services for the floating population. We should make basic medical and health services more accessible for migrant workers and their children. Migrant workers' children who live with their parents can receive the same basic medical and health services as children with household registration in the place of residence.

We should improve public health services and the prevention and control of major infectious diseases for the floating population and equalize public health services between migrant workers and urban residents. We should better ensure food safety, advance the supporting legislation for the Food Safety Law, and improve the food safety work mechanism. We should strengthen the food safety risk-monitoring network, integrate monitoring resources, establish a unified national food safety risk-monitoring system, and improve the food safety risk communication system. We should strengthen food safety standard setting and revision and complete the collation and integration of existing food safety standards as soon as possible. We should strengthen standard setting and revision for key food specialties and segments and enrich and improve the national food safety standard system. We should establish sound food safety accident information reporting and epidemiological investigation mechanisms and improve the food-borne disease monitoring and emergency response capabilities of disease prevention and control institutions at all levels.

We should intensify our efforts in maternal and child health. We should strengthen municipal and county maternal and child healthcare institutions. We should better prevent and treat common gynecological diseases in rural areas, focusing on cervical cancer and breast cancer screening. Our aim is that in 2030, more than 90% rural women who should be examined for common diseases will actually get examinations. We should establish emergency centers and green channels for critically ill pregnant women and newborns to improve the quality of obstetrics and pediatrics services. We should increase the intervention on birth defects with three-level comprehensive prevention and treatment of birth defects. We should strengthen communication and education on premarital and prepregnancy healthcare, prenatal screening and prenatal diagnosis, and management of newborn disease screening. We should comprehensively manage pediatric diseases and other appropriate technologies and, particularly, improve the ability of rural medical and health institutions for common pediatric disease diagnosis and treatment, on-site first aid, and the treatment and referral of critically ill children.

We should improve the health supervision system. We should enhance the grassroots health supervision network and monitoring and improve the monitoring network' direct reporting system. We should strictly manage health supervision teams and regulate law enforcement. We should carry out quantitative and hierarchical health supervision management for public places. We should continue the special supervision and sampling inspections over disinfection products and products concerning drinking water hygiene and safety. We should better supervise and manage radiation hygiene focusing on radiation hazards in medical institutions. We should further monitor and evaluate environmental pollution's health impact and better monitor, diagnose, and treat heavy metal pollution's health hazards. We should carry out comprehensive school health supervision, focusing on rural and other vulnerable areas. We should tighten the supervision and inspection for infectious disease prevention and control. Crackdown on illegal medical practice and blood collection and supply should be intensified.

We should improve the monitoring system for key endemic diseases, take relevant prevention and control measures, and basically eliminate the harm of key endemic diseases. We should adhere to comprehensive iodine deficiency prevention and control measures represented by iodized salt. We should fully optimize water sources, boilers, and stoves in endemic fluoride and arsenic poisoning areas. We should practice integrated management of rabies, brucellosis, and other zoonotic diseases focusing on source control. Among them, we should reduce the mortality rate of rabies and curb the rising of brucellosis. We should strengthen the comprehensive prevention and control of the hand-foot-and-mouth disease, and influenza monitoring and prevention.

4.5.4 *Improving the Medical Security System*

We should establish a sound multilevel medical security system for urban and rural residents. We should lift basic medical insurance' subsidy amount, payment cap, and payment ratio. We should consolidate and expand the system and elevate its security level. We should narrow the gap between urban and rural areas in medical insurance financing and security to support a unified urban and rural medical security system.

We should optimize the urban and rural medical assistance system, increase medical assistance in all aspect, and subsidize the recipients for insurance premium and medical expenses they cannot afford, to provide them with a solid safety net. We should seek a security mechanism for major and extraordinarily serious diseases. We should provide critical illness insurance for urban and rural residents and use the basic medical insurance fund to purchase commercial critical illness insurance to relieve the burden of high medical expenses for the insured. We should let basic medical and critical illness insurances, medical assistance, supplementary insurances, public welfare, and charity work in synergy and mutual complement. We should coordinate basic medical, critical illness, and commercial health insurance policies for effective security.

We should comprehensively advance the reform of payment methods. Based on fund revenue and expenditure budget management and disease clinical channel management, we should promote a variety of payment methods across the nation, including disease-specific payment, capitation, and global budget. We should establish a restrictive mechanism for medical insurance to constrain excessive medical expenses.

We should improve the supplementary medical insurance system by developing commercial health insurance. We should develop more policies for the commercial health insurance industry and encourage commercial insurers to develop health insurance products other than basic medical insurance to meet diverse health needs. We should encourage enterprises and individuals to be enrolled in commercial health insurance and supplementary insurances.

4.5.5 *Establishing a Sound Drug Supply Assurance System*

We should revise and amend the national drug policy according to the development of the pharmaceutical industry. The policy is a comprehensive and common framework to guide stakeholders in the pharmaceutical industry. It consists of many objectives and measures, including drug development policy, production supply policy, use policy, and economic policy. The national drug policy has strong practical significance for solving current prominent problems in China's medical and health field. It can play a vital role in ensuring the availability and rational and safe use of essential medicines for Chinese people.

We should consolidate similar drug lists and implement centralized management. There are many benefits of doing so. First, it streamlines organizational organs and saves management costs. Second, once carried out, the reform of payment methods based on the consolidated list can encourage hospitals to use low-price drugs. Third, it reduces the number of enterprise bids and saves social costs. Fourth, it is conducive to medical insurance institutions to play a supervisory role and avoid waste. At the same time, we should actively advance the reform of payment methods, implement global budget, capitation, or disease-specific payment, and encourage hospitals to take the initiative to use low-price medicines. We should actively promote the site-off settlement of medical insurance to facilitate the floating population in using medical insurance. We should seek a drug procurement method that adapts to medical insurance payment methods, expand the autonomy of hospitals in drug procurement, and break the existing buyer's monopoly.

We should reform the drug pricing mechanism and form a drug pricing committee. Commodity price authorities and the expert advisory committee should jointly decide the initial price, which will be discussed and heard by the pricing committee. The discussion and hearing should be open to all parties, so as to establish a mechanism for ensuring science-based pricing. Drawing on advanced experience from other countries, we should introduce pharmaco-economics into the drug pricing mechanism. We should price generic drugs and innovative drugs differently to encourage enterprises to develop and produce drugs with higher clinical value. We should implement a clinician reporting system. Doctors' sentiments and research on the use of drugs in the clinical process, such as the scope of indications, the actual incidence of side effects, and the degree of improvement of the patient's quality of life, will be reference for drug pricing. We should timely disclose cost information review, drug price approval and negotiation, and information on other aspects. The pricing must have sufficient basis and be open, transparent, and subject to social supervision.

We should establish a sound drug cost monitoring system to monitor costs and use it to establish an objective, scientific, and reasonable joint cost and price survey mechanism. We should monitor changes in the cost, supply, and demand of active pharmaceutical ingredients, as well as technical difficulties and changes in drug production, and grasp all legitimate expenditures in drug circulation, so as to support the final calculation of the average social cost of drugs. We should also account for the cost of bringing innovative medicines to market. Based on the cost reported by the enterprise and the information collected, we will commission experts in pharmaceutical and price research to evaluate and calculate the average social cost of drugs.

We should reform and improve China's drug procurement system. We should allow the market to play the foundation role in resource allocation more extensively and deeply to improve drug procurement efficiency. We should redesign the existing centralized drug procurement system and coordinate the roles of the government and the market in drug procurement. We should improve the functions of online drug procurement platforms so that the whole procurement process can be completed using information technology. Ultimately, we will establish a drug procurement information platform that integrates the producer's qualifications, supply information, and hospital drug procurement. We should comprehensively evaluate the efficacy, quality, and price of drugs, and set science-based evaluation indicators. At the same time, we should establish a sound supervision system for designing collective decision-making systems for the selection of varieties, bid evaluation, and bid awarding.

We should strengthen the whole-process quality supervision on drug development, production, circulation, and use. We should implement the National Drug Standard Improvement Action Plan to comprehensively improve the quality of generic drugs. We should optimize the drug inspection and testing system and improve the inspection and testing capabilities. We should strengthen the grassroots monitoring of adverse drug reactions and the evaluation and early warning of adverse drug reactions and medical device adverse events. We should optimize the drug safety emergency response system and improve the emergency response capability and caliber. We should promote the construction of the national electronic supervision system for drugs, and improve the electronic supervision system for drugs covering all varieties and the whole process and with traceability.

We should further improve the traditional Chinese medicine (TCM) service system and strengthen county TCM hospitals. We should carry out TCM-based prevention, control, and research on major diseases. We should intensify the preservation, research and development, and rational exploitation of TCM resources and boost the Chinese herb industry. We should cultivate high-caliber TCM talents and foster TCM masters. We should seek a TCM inheritance and innovation system. We should intensify the inheritance and development of ethnic medicine and integrate traditional Chinese with modern medicine. We should advance the legislation, IT-based transformation, and standardization of TCM. We should develop TCM culture and promote TCM to the world. We should study and develop medical security and essential medicine policies that encourage TCM services, and improve the security mechanism for TCM development.

We should vigorously develop biomedicine and transform and upgrade traditional medicine. We should revise policies for the pharmaceutical industry, encourage the merger between and reorganization of pharmaceutical enterprises to increase industrial concentration. We should support enterprises to accelerate technological transformation and enhance the industry's core competitiveness and sustainability . We should strengthen independent innovation and comprehensively improve biopharmaceutical enterprises' innovation capability and product quality management. We should encourage the development, commercialization, and application of new products and processes such as biotechnology drugs, chemical drugs, Chinese herb, and biomedical engineering. We should vigorously develop TCM-related health industries and encourage and support the industry-university-research integration and the establishment of industrial technology alliances so as to improve the international competitiveness of China's Chinese herb industry.

4.5.6 *Developing a Quality Grassroots Medical and Health Workforce*

We should speed up the roll-out of the strategy to develop a high-quality health workforce and vigorously improve the medical and health training system and mechanism innovation. We should bolster the development of a grassroots medical and health workforce focusing on general practitioners. We should establish a training network with clinical training centers and grassroots practice centers as the main body, which focuses on standardizing and improving clinical diagnosis and treatment capabilities and public health service capabilities. We should improve the comprehensive medical training system, integrating college education, postgraduate education, and further education.

We should adjust the distribution of health workforce in urban and rural areas. We should train more health workers for rural areas, provide free education for medical students appointed to rural areas, and train leading doctors for county hospitals. We should provide further education and practical skill training for grassroots medical and health workers. We should develop preferential policies to encourage and guide medical workers to work at the grassroots institutions. We should strengthen the training of village health workers, and turn village doctors into physicians and physician assistants. We should study and implement plans for general practitioners at grassroots medical and health institutions and special positions for high-caliber talents at county hospitals in urgent need. We should strengthen the development of public health workforce and improve the position management system of professional public health institutions to attract and encourage outstanding talents to engage in public health. We should establish a sound standardized training system for public health physicians.

We should strengthen the research on the current medical service demand and supply system in order to clarify the development orientation for the medical service system. We should improve the health human resources reserve mechanism according to the development needs of health services and develop and implement visionary medical student training plans. We should bring in more pediatricians, psychiatrists, and nurses, which are in shortage. We should seek an open training and practice system for nurses, as well as professional institutions for nurse training and talent pools. We should encourage eligible nurses to set up nursing facilities to provide rehabilitation nursing services as a way to activate nursing resources. We should speed up the development of new workforces such as psychotherapists, rehabilitation therapists, and medical social workers. We should further improve the standardized training of residents and general practitioners, and establish a standardized training system for specialist physicians and public health physicians. We should encourage more workers to the grassroots level and qualified practitioners to serve as family doctors on a full-time or part-time basis.

We should set short-term and long-term goals for the reform of China's medical worker compensation system and prioritize ensuring and increasing the wages of grassroots medical workers. We should develop promotion plans for outstanding medical and health experts by category. We should establish a system for professional health administrative staff and ensure all health administrative staff are specialized and professional workers. We should innovate the training, use evaluation, flow allocation, and incentive assurance mechanisms for medical and health personnel, and provide more preferential policies for the development of medical and health personnel.

4.5.7 *Promoting the IT Transition of Medicine and Health Undertakings*

We should strengthen regional information platforms and build a national information platform. We should speed up the establishment of a nationally unified standard system for medical and health information technology. We should build regionally unified appointment and registration platforms and implement diagnosis and treatment appointments generally. Electronic medical records should be shared among medical institutions between regions. We should coordinate the management of health statistics, epidemic reports, and information of health supervision, medical treatment, and medical service supervision, shifting from individual management to real-time supervision and comprehensive management. We should guide and promote commercialized medical and health information services. We should gradually realize the information connectivity among the application systems for medical services, public health, medical security, drug supply assurance, and comprehensive management, as well as medical and health information sharing.

We should speed up the construction of the residents' health record and e-medical record information systems. We should apply health big data more widely and allow the population information, electronic health record, and electronic medical record databases basically cover the whole population as soon as possible, with timely information update. We should make full use of information technology to improve residents' health and chronic disease management and increase the coverage of electronic health records for urban and rural residents. We should provide people with continuous prevention, healthcare, medical treatment, and rehabilitation services to facilitate residents' participation in personal health management.

We should strengthen the IT transition of medical institutions. We should establish information systems for hospital diagnosis and treatment practice management and medical personnel performance evaluation to standardize medical service practice and improve resource utilization efficiency. We should speed up the IT transition of grassroots medical and health institutions. We should establish provincial grassroots medical and health information systems that feature functions such as the supply and use of essential medicines, residents' health management, basic medical services, and performance appraisal.

We should apply new technologies and models such as mobile Internet, Internet of Things, cloud computing, telemedicine, and wearable devices to promote health information services and smart medical services that benefit all. We should support the development of new diagnosis and treatment models such as telemedicine and mobile medical treatment for rural and remote areas and enhance the efficiency and catalytic role of high-quality medical resources, so as to make it more convenient for patients to seek medical treatment and improve and equalize grassroots medical and health services especially in remote areas.

CHAPTER 5

Modernization of Basic Elderly Care Services

5.1 Overview of Basic Elderly Care Services

5.1.1 *Definition of Basic Elderly Care Services*

Elderly care service is a general term for services to meet the elderly's special needs. It is a definition from the perspective of life course, referring to services to meet the elderly's special needs in physical, psychological, and other aspects due to aging, diseases, or other reasons. Elderly care services mainly include daily care, medical care, rehabilitation care, and spiritual and cultural services. Elderly care services can be divided into basic and non-basic elderly care services.

Basic elderly care services are provided by the government according to the country's economic and social development and secure life necessities for all senior citizens. Non-basic elderly care services are paid services provided by the private sector to improve the quality of life (QOL) of senior citizens with effective demand, which are beyond the scope of basic elderly care services or above the basic level.

Basic elderly care services are characterized as follows. First, the government leads the provision of basic elderly care services. Basic elderly care services have significant positive externalities and entail market failures. Therefore, it is an important function of the government to ensure the provision. Second, basic elderly care services are basic and sustainable services provided within a designated scope and at an appropriate level to

K. Dong, N. Wei, *Transforming China's Public Services*,
https://doi.org/10.1007/978-981-99-3943-5_5

ensure senior citizens' life necessities. The services' quantity and level should be within the financial capacity of the state so that it can be sustainable. The content and standards of basic elderly care services are, of course, not static. They will be adjusted according to economic and social development. Finally, basic elderly care services are provided by the state for senior citizens in accordance with the Constitution and other applicable laws, and all senior citizens have the right to such services according to law.

5.1.2 Details of the Basic Elderly Care Service System

The basic elderly care service system is a network of facilities, organizations, the workforce, and technical elements for elderly daily care, rehabilitation care, and other services, aiming to meet their basic elderly care service needs and improve their QOL, which is commensurate with economic and social development. It is supported by specific service standards, operating mechanisms, and regulatory systems. China's basic elderly care service system consists of three parts: home-based care services, community-based care services, and institution-based care services.

Home-Based Care Services[1]

Home-based care services are mostly door-to-door services for seniors living at home. They include daily care, housekeeping, medical care, rehabilitation care, and spiritual comfort for the elderly. Home-based care services are complicated. First, they are for seniors living at home. Second, funds for home-based care services may come from the family (children, spouses, relatives, or seniors themselves), the government, and society (social pensions, social welfare, social assistance, and charity), representing the characteristics of sharing among the individual, the family, the government, and society. Third, home-based care services may be provided by family members (children, spouses, and relatives), the community, and the private sector (non-governmental non-profit institutions and intermediary organizations, for-profit enterprises, and elderly care service institutions).

[1] Yan Qingchun, *Program of Action for the Development of Undertakings for the Aged in the New Era: An interpretation of the Plan for Development of China's Undertakings for the Aged During the 12th Five-Year Plan* (Beijing: China Society Publishing House, 2012): 135.

Fourth, home-based care services include daily care, medical care, health management, cultural education, and spiritual comfort, among other aspects. Fifth, home-based care services include non-profit and public welfare services as well as for-profit services.

Community-Based Care Services

"Community-based care services are an important support for home-based care services. They have two functions: community day care and support for home-based care. They are mainly for seniors in the community who live alone or their family members cannot care for them during the day."[2] Community-based care services provide the seniors with convenient and professional services in the community which they live in and are familiar with, combining the advantages and characteristics of home- and institution-based elderly care. In terms of the form, community-based care services provide the seniors with a variety of door-to-door services, as well as day care services through platforms such as community day care centers.

Institution-Based Care Services

For institution-based care services, the state, social organizations, or individuals establish elderly care institutions to provide the elderly with centralized residence, daily care, rehabilitation care, cultural entertainment, spiritual comfort, and other services. Institution-based care services are primarily for seniors living in dire poverty who need to be supported by the government, seniors with disabilities or semi-disabilities whose family members are unable to take care of them, people over 80 and empty-nesters, and well-off seniors who are willing to be taken care of at elderly care institutions. Institution-based care services are an important vehicle of basic elderly care services, which can effectively meet the elderly's needs for centralized services and provide social elderly care services for the elderly in need, especially those with disabilities or semi-disabilities. Being highly professional, institution-based care services can support home- and community-based care services in nearby communities and improve the overall level of social elderly care services.

[2] www.gov.cn/xxgk/pub/govpublic/mrlm/201112/t20111227_64699.html.

5.1.3 *Development of Basic Elderly Care Services Vital to National Economy and People's Livelihood*

In the end of 2015, China's elderly population reached 222 million. As those born in the second baby boom from 1962 to 1976 are aging, the period from 2020 to 2030 will be the fastest population aging period in China. In 2020 and 2030, China's elderly population will reach 255 million and 371 million, respectively, with an average annual net increase of 11.59 million. In terms of proportion, the percentage of China's elderly population will increase from 16.1% in 2015 to 17.8% in 2020 and 25.3% in 2030.[3] With the rapid population aging, the population of people over 80, empty-nesters, seniors with disabilities will increase drastically. The population of people over 80 in China is expected to reach 29.2 million in 2020 and 43.4 million in 2030, with an average annual increase of 1.42 million. In 2015, the proportion of empty-nesters in China exceeded 50%. With the development of economy and society, this figure is expected to continue to rise from 2020 to 2030. Seniors with disabilities or semi-disabilities are over 40 million in 2015, accounting for nearly 20% of the total elderly population. From 2020 to 2030, their number will increase rapidly. The huge and fast-growing number of people over 80, empty-nesters, and seniors with disabilities will inevitably create an enormous demand for elderly care services, which will become a basic reality that China must respond to in its economic and social development. Chinese families are significantly downsizing, and home-based elderly care is challenged by insufficient resources and family members' declining willingness for elderly care. These factors will lead to a surge in demand for basic elderly care services in China. Therefore, vigorously developing basic elderly care services as an active response to the huge demand has become the only choice to meet the elderly's livelihood needs.

Vigorously developing basic elderly care services is vital to expanding domestic demand and promoting economic growth. With the rise of global trade protectionism, exports have encountered more and more frictions, barriers, and obstacles in expansion, and international competition has become more intense. At the same time, large-scale government investment is unsustainable due to disadvantages such as excessive production capacity and weakening social investment. Lessons from other

[3] General Research Team of China on Strategic Response to Ageing Population, *General Report of China on Strategic Response to Ageing Population* (Beijing: Hualing Publishing House, 2014): 191–192.

countries show that for both developed and developing countries, national economic development mainly depends on domestic demand. Only by expanding domestic demand can a country take an invincible position in development. Vigorously developing basic elderly care services can effectively stimulate consumption among the elderly as well as elderly care workers, thus effectively expanding domestic demand.

In addition, vigorously developing basic elderly care services is conducive to opening up new employment sectors and creating jobs. For the present and a long period ahead, employment pressure cities will increase, and the transfer of surplus rural labor to non-agricultural fields will accelerate. As a result, China will face arduous tasks of ensuring employment. Elderly care services are a labor-intensive industry with prominent ability to absorb labor. At a low ratio of 1:10 (one caregiver for every 10 seniors with disabilities), the number of caregivers needed by seniors with disabilities alone is expected to increase from 4 million in 2015 to 9.75 million in 2050. Therefore, vigorously developing basic elderly care services and other elderly care services can effectively expand employment sectors and increase the total number of jobs.

5.2 Remarkable Progress Made in Building China's Basic Elderly Care Service System

5.2.1 Establishing a Sound Basic Elderly Care Service System Becoming a Major Strategic Measure to Actively Respond to an Ageing Population

The Chinese government attaches great importance to undertakings for the aged. As population aging intensifies, the Chinese government has issued many major policies and regarded a sound basic elderly care service system as a major strategic measure and an actively response. The Plan for Development of China's Undertakings for the Aged During the 12th Five-Year Plan proposes to "establish and improve the social elderly care service system which is based on families and supported by communities and institutions." The Plan for the Development of the National Basic Public Service System During the 12th Five-Year Plan Period points out that it is necessary to "establish a sound elderly care service system, encourage home-based elderly care, expand communities' functions of elderly care services, and enhance the service capabilities of non-profit elderly care service institutions." The Law of the People's Republic of China on the Protection of the Rights and Interests of the Elderly stipulates, "The state shall establish and improve the social elderly care service system which is based on families and supported by communities and institutions." The Several Opinions on Strengthening Development of the Elderly Care Service Industry set a clear development goal that "by 2020, a comprehensive elderly care service system with complete functions, a moderate scale, and coverage of urban and rural areas, which is based on families and supported by communities and institutions, will be established." The Outline of the 13th Five-Year Plan for the National Economic and Social Development of the People's Republic of China (the "Outline of the 13th Five-Year Plan") proposes to "build a multi-level elderly care service system which is based on families and supported by communities and institutions."

5.2.2 *The Law and Policy System for Basic Elderly Care Services Coming into Shape*

In recent years, China has stepped up efforts in the legislature for basic elderly care services and introduced many major policies on basic elderly care services. The top-level design has been improved and a system of laws and policies for basic elderly care services has come into shape.

In terms of comprehensive policies, the newly revised Law of the People's Republic of China on the Protection of the Rights and Interests of the Elderly has been promulgated and enforced. The State Council has issued some landmark policies, including Several Opinions on Strengthening Development of the Elderly Care Service Industry and Several Opinions on Promoting the Development of the Health Service Industry.

In terms of planning, the State Council has issued the Plan for the Development of the National Basic Public Service System During the 12th Five-Year Plan Period, the Plan for the Development of the Services Industry During the 12th Five-Year Plan, and the Plan for the Development of China's Undertakings for the Aged During the 12th Five-Year Plan. The Ministry of Civil Affairs has issued the Plan for Social Elderly Care Service System Construction (2011–2015) and the 12th Five-Year Plan for the Development of Civil Affairs. The General Office of the State Council has issued the Plan for Construction of the Service System in Communities (2011–2015).

Major policies concerning management and construction include: the Measures for Licensing for Establishing Elderly Care Institutions, the Measures for the Administration of Elderly Care Institutions, the Basic Code for Elderly Care Institutions, the Monitoring Indicator System for the Development of Social Elderly Care Services, Opinions on Comprehensively Promoting Home-based Care Services, the Guiding Opinions on Promoting the Evaluation on Elderly Care Services, Opinions on Establishing Elderly Services Cooperation and Counterpart Support Mechanisms, the Notice on Relevant Issues Concerning Reduction and Exemption of Administrative Charges on Elderly Care Institutions and Medical Institutions, the Guiding Opinions on Regulating the Administration of Service Charges of Elderly Care Institutions to Promote the Healthy Development of the Elderly Care Service Industry, Opinions on Promoting Liability Insurance for Elderly Care Institutions, and the Notice on Effectively Handling Government Purchases of Pension Services.

Major policies concerning standard setting include: the Guiding Opinions on Strengthening the Standardization of Elderly Services, the Design Code for Buildings of Elderly Facilities, the Construction Standards for Elderly Nursing Homes, the Construction Standards for Community Day-care Centers for the Elderly, the Construction Standards for Urban Community Service Stations, the Basic Standards for Clinics in Elderly Care Institutions (for Trial Implementation), and the Basic Standards for Nursing Stations (for Trial Implementation).

Major policies concerning reform include: the Notice on Launching Reform Pilot Work for Government-run Institutions for the Elderly and the Notice on Carrying Out the Comprehensive Reform Pilot Program for the Elderly Care Service Industry.

Major policies concerning personnel training include: Opinions on Accelerating Personnel Training for the Elderly Care Service Industry, the National Occupation Standards for Senior Caregivers, and the Notice on Selecting Demonstration Programs of Elderly Care Services in Vocational College in China.

Major policies concerning facility construction and renovation include: the Notice on Enhancing the Planning and Construction of Elderly Care Service Facilities, the Notice on Enhancing the Construction of Accessible Public Facilities in the Elderly's Homes and Residential Areas, the Notice on Promoting the Construction of Urban Pension Service Facilities, and the Guiding Opinions on Land-Use for Elderly Care Facilities.

Major policies concerning fund raising and allocation for basic elder care services include: Opinions on Establishing a Unified Basic Pension Insurance System for Urban and Rural Residents, the Measures for the Administration of Medical Assistance Funds in Urban and Rural Areas, the Interim Measures for the Integration of Urban and Rural Pension Insurance Systems, and the Management Measures for the Central Special Lottery Public Welfare Fund Supporting Rural Nursing Homes.

Major policies concerning promoting the diversified development of basic elder care services include: the Implementation Opinions on Encouraging and Guiding Private Capital to Enter the Field of Elder Care Services, the Implementation Opinions on Encouraging Private Capital to Participate in the Development of the Elderly Care Service Industry, the Notice on Issues Concerning the Establishment of For-profit Elderly Care Institutions and Disability Service Institutions in the Mainland by Service Providers from Hong Kong and Macao, and the Announcement on Encouraging Foreign Investors' Establishment of For-Profit Elderly Care Institutions in China to Provide Elderly Care Services.

Major policies concerning the integration of basic elder care services and medical care and health services include: Several Opinions on Promoting the Development of the Health Service Industry, Several Opinions on Encouraging and Guiding the Healthy Development of Private Investment, the Notice on Accelerating Development of Health and Pension Service Projects, the Notice on Organizing and Carrying Out the Pilot Program for the Policy of Telemedicine for Elderly Care Institutions, and the Standard of Health Management Service of TCM: TCM Health Management Service for the Elderly.

5.2.3 *The Basic Elderly Care System in Rapid Advancement*

Home-based care services started pilot programs in some Chinese provinces (autonomous regions and municipalities directly under the Central Government) in 2000. The services have been rolled out to varying degrees in large and medium-sized cities in most provinces (autonomous regions and municipalities directly under the Central Government) across the country. In the end of the 12th Five-Year Plan, home-based care service facilities have covered most urban communities and more than 50% of rural communities in China. On the whole, home-based care services have a preliminary organizational system, a functioning subsidy system, and a considerable service workforce. A supervision mechanism is to be established.

In terms of community-based care services, nursing homes, day care centers, elderly rehabilitation centers, and other community facilities have been established in sub-districts and communities in cities, providing a variety of convenience services for the elderly. The construction of comprehensive community service facilities has been under steady progress. In December 2015, there were 81,000 urban community comprehensive service facilities nationwide, with a coverage rate of 82.12%. Remarkable results have been achieved in building a community service workforce. IT application in communities has been promoted as scheduled. The system and mechanism of community services have been undergone continuous innovation.[4]

In terms of institution-based care services, in the end of 2015, the number of elderly care service beds reached 6.698 million, and the number of elderly care service beds per 1000 elderly people was 30.3, an increase of 70.3% over the end of 2010.[5] Private capital has been enthusiastic about investing in infrastructure construction, and the private sector has become the main actor to build elderly care institutions. The restructuring of public elderly care institutions has been officially launched, and the market-oriented operation reform has been accelerated. Elderly care institutions are more diverse and their services see improvements.

New progress has been made in basic elderly care services in rural areas. Many regions are trying out "home-nursing-house integration" and other elderly care service models that meet the elderly's needs in rural areas. Practices such as rural mutual-aid nursing homes in Hebei and Jilin and care service centers for people with long-term illness from rural families in Liaoning have been widely promoted. With the continuous expansion of rural community construction, rural communities have seen improvements in their functions of elderly care services.

5.2.4 A Basic Elderly Care Service Management System and Operation Mechanism Led by the Government and Involving Various Participants Taking Shape

The government has strengthened organizational leadership. It conscientiously performs functions such as planning guidance, policy formulation, specification setting, supervision, and management. Coupled with the

[4] http://ftzb.mca.gov.cn/asop/templates/n52/interviewlog.html.
[5] http://ftzb.mca.gov.cn/asop/templates/n52/interviewlog.html.

close cooperation of other participants, a joint force has been formed to develop basic elderly care services. This is mainly manifested in: First, governments at all levels have issued many development plans, policies, systems, and standards aimed at promoting the development of basic elderly care services. Second, the government has organized or coordinated to establish basic elderly care service management systems and service organizations and form a basic elderly care service network. Third, a number of basic elderly service infrastructure facilities have been established using government financial input. During the 12th Five-Year Plan period, governments at all levels increased investment in the construction of basic elderly care service facilities. The investment in the central budget alone reached CN¥10.8 billion, and the lottery public welfare fund at all levels was CN¥28.5 billion.[6] Fourth, under the guidance of the government, private capital has been encouraged and guided to enter the field of elderly care services through various means such as policy support, financial subsidies, private running with public subsidies, public constructing, and private running. Fifth, more effort has been put into the construction of elderly care service industry organizations; a management system comprised of the government, intermediary organizations, and elderly care service institutions has been established; and industry self-discipline, management, exchanges, and training have been carried out.

5.2.5 *Marketization of Basic Elderly Care Services Accelerated*

With the development of the government's governing view and the transformation of government functions, the market plays a more important role and is more valued in the development of basic elderly care services. In particular, since the 18th CPC National Congress, the marketization of basic elderly care services has been accelerated. The 18th CPC National Congress reaffirmed to "leverage to a greater extent and in a wider scope the basic role of the market in allocating resources." The third plenary session of the 18th CPC Central Committee further emphasized "enabling the market to play a decisive role in resource allocation." From "the basic role" to "a decisive role," it indicates the CPC Central Committee's emphasis on the role of the market and its strong willingness to further promote marketization. Several Opinions on Strengthening Development of the Elderly Care Service Industry and Several Opinions on Promoting

[6] http://ftzb.mca.gov.cn/asop/templates/n52/interviewlog.html.

the Development of the Health Service Industry issued in 2013 and the Outline of the 13th Five-Year Plan released in 2016 have made it a basic principle and main task to improve the market mechanism, give full play to the role of the market, and make the private sector the main actor in developing the elderly care service industry and health service industry. At the same time, local governments at all levels have introduced preferential policies and measures to make basic elderly care services appeal to investment. The private sector has poured in, and the marketization of basic elderly care services has been advanced rapidly.

5.2.6 *Increasing Sources and Diversified Allocation Methods of Funds for Basic Elderly Care Services*

With the development of China's economy and society, the raising and input of funds for basic elderly care service system construction have shifted from the model where "the government, the entity, and the rural collective take their own responsibility and operate in an isolated manner" to the model where "the government, the market, and society collaborate with diversified investment." This transformation is mainly manifested in three aspects. First, governments at all levels have increased their investment in basic elderly care service system construction. Second, enterprises participate in basic elderly care service system construction through a market-oriented mechanism. Third, non-profit elderly care service projects organized by social organizations and charitable donations to basic elderly care services from all sectors have increased. At the same time, with the increase of investors, operators, and operating models of basic elderly care service projects, the models of capital allocation, or service provision, have been more diversified. First, in addition to meeting the basic elderly care service needs of the elderly supported by the government, public-funded elderly care institutions are receiving seniors who are willing to pay to live in them. Government-funded community service organizations provide free or paid home-based or community-based services for seniors. Second, most private-funded elderly care service projects provide different levels of paid services to different types of seniors. Third, elderly care service projects funded by social organizations provide various free, affordable, and paid services to different types of seniors. Fourth, the government, the market, and society have strengthened cooperation in elderly care service projects, and new models such as public constructing and private running, private running with public subsidies, financial subsidies, and government procurement have emerged. It not only ensures the social benefits of basic elderly care service projects but also improves their operation efficiency.

5.3 Prominent Problems Remaining in China's Basic Elderly Care Service System

In recent years, especially since the 18th CPC National Congress, efforts have been intensified in China's basic elderly care service system, which have brought remarkable progress. The elderly has gained more and been happier. However, due to backward mindset and some negative legacies, there are still many problems that need to be solved urgently in China's basic elderly care service system.

5.3.1 Misallocation of Certain Basic Elderly Care Services

Some basic securities are absent in basic elderly care services. Government investment projects are mistakenly equated with basic elderly care services, resulting in excessively high standards. For example, in institution-based elderly care services, public elderly care institutions shall provide basic elderly care services as assistance and living necessities. They shall ensure the basic life of senior residents, instead of providing services that are higher than the local average living standard. Therefore, the construction standards, service quality, and grades of public elderly care institutions shall be benchmarked with basic and general elderly care services. However, in reality, many local governments regard public elderly care institutions as an image project and continuously increase their financial investment. It has even been common to build luxury demonstration public elderly care institutions, with facilities and services much higher than the level of basic needs. On the other hand, the public elderly care institutions in some remote areas have very simple facilities and poor services, where senior residents cannot access even the most basic services.

5.3.2 Prominent Contradiction Between Supply and Demand

China has just turned an aging society, and basic elderly care services have just started to develop. In many aspects, it is still in the stage of exploration and reform. There are many contradictions between the supply and demand of basic elderly care services, which are mainly manifested in insufficient total volume, unreasonable structure, limited varieties, and backward services.

In terms of home-based care services, many social organizations and enterprises lack the necessary surveys and understanding of the elderly's needs for home-based care services. They only provide limited services based on their assumption, most of which are housekeeping services, and fail to meet the elderly's actual needs. In addition, the current home-based care services are only for seniors with low-income and disabilities in urban areas who access such services through government procurement of services.

Community-based care services as a whole are primarily community day care. However, due to backward guiding concepts, many community day care centers have become recreation centers for seniors able to take care of themselves. Most of their beds are vacant. Seniors who are truly in need rarely have access to day care services.

In terms of institution-based care services, it is difficult to find a bed in some of the existing elderly care institutions, while a large number of beds are idle in others. It indicates a prominent structural problem. For example, in rural areas, the elderly has low demand for elderly care institutions but there are a large number of elderly care institutions and beds. In urban areas, elderly care institutions are in high demand but most of them are located in suburbs farther from the downtown. The elderly has high demand for nursing elderly care institutions but most of the institutions are for seniors able to take care of themselves.

5.3.3 Imbalanced Development

Basic elderly care services are imbalanced between regions, between urban and rural areas, and between sectors. Basic elderly care services in the developed eastern regions are far more developed than those in the central and western regions. On the other hand, the aging of the central and western regions is and will be faster than that in the eastern regions, which poses a huge challenge to the basic elderly care service system in the central and western regions. Basic elderly care services in rural China are far behind those in cities in terms of infrastructure construction, investment, overall caliber of service workforce, service content, and service quality. In the face of faster population aging, the development of the rural elderly care service system needs to be accelerated. In addition, basic elderly care service-related sectors vary greatly in development. Specifically, the elderly nursing and health service sectors are growing rapidly, while the elderly culture, sports, and entertainment industries have lagged behind.

5.3.4 *Institution-Based Care Services Overemphasized While Home- and Community-Based Care Services Neglected*

The Chinese government has made it clear that the basic elderly care service system shall be "based on families and supported by communities and institutions." Following this instruction, some local governments have envisaged the "90-6-4" and "90-7-3" (percentage of seniors receiving elderly care at home, in the community, and at the institution) elderly care patterns. However, in reality, home- and community-based care services develop far slower than institution-based care services. It is mainly because that building an elderly care institution is a "visible measure with quick returns," coupled with the bad tendency of "image project." In fact, seniors can access services they need in a timely and convenient manner at home or in the familiar community, which better conforms to traditional Chinese elderly care. They cost less and are proven and recommended by the international community. Therefore, in constructing the basic elderly care service system, China must distinguish between the primary and the secondary and put effort in a targeted manner.

5.3.5 *Improvement Needed for the Government's Macro-Management*

There are many administrations and funding sources for China's undertakings for the aged but lacks a strong coordinator. Some of the relevant agencies work in a separate system, while others are affiliated to the civil affairs authorities or other departments. It makes it difficult for them to perform "comprehensive coordination, supervision and inspection, and offering advice." As a result, China's basic elderly care service management is divided by departments, regions, and institutions of different ownership. The lack of necessary coordination between departments makes it difficult to form a synergy for elderly care services. In particular, private elderly care services urgently need strong sector guidance and management. Many elderly care service institutions see low occupancy rates and serious waste of resources.

5.3.6 *Workforce Lagging Behind*

Human resources are crucial China's basic elderly care services. However, through recent years, it has always been a weak point. Specifically, professional managers are very scarce, service workers are of greater ages and have poor professional skills, and there is a large shortage of volunteers. Take senior caregivers as an example. At present, most elderly care service workers in China are the middle-aged, unemployed people, and migrant workers. Only tens of thousands of people obtain the national vocational qualification certificate for senior caregiver every year, and there are less than one million workers nationwide. Many caregivers have only junior high school or even primary school education, and nearly none of them have undergone systematic training or basic knowledge of elderly care. They can only cope with daily care and housekeeping services but not the elderly's diverse needs at different levels and with different characteristics. Few elderly care service institutions are equipped with nursing, psychological counseling, legal services, or social work professionals.

5.4 Actively Advancing the Construction of China's Basic Elderly Care Service System from 2020 to 2030

5.4.1 *Fundamental Basis*

Economic and Social Development Strategies
According to the reports of the 15th and 16th CPC National Congresses, after China becomes a moderately prosperous society at the end of the twentieth century, it will achieve modernization by 2050 in three stages divided by 2010 and 2020. The first stage will be completed in 2010, the second will be from 2010 to 2020, and the third will be from 2020 to 2050, when China will basically achieve modernization.

The report of the 18th CPC National Congress instructs to "fully implement the overall plan for promoting economic, political, cultural, social, and ecological progress" to "attain the goal of completing the building of a moderately prosperous society in all respects by 2020" and "generally achieve equal access to basic public services." The Outline of the 13th Five-Year Plan further called on "building a multi-level elderly care service system which is based on families and supported by communities and institutions" and that "public service systems will be improved, with access to basic public services becoming increasingly equitable."

The basic elderly care service system is an important part of economic and social development, and its development must follow and serve the overall strategy of economic and social development.

Development Plan for Basic Public Services
The Plan for the Development of the National Basic Public Service System During the 12th Five-Year Plan Period and the Plan for the Development of the Services Industry During the 12th Five-Year Plan have made overall arrangements and planning for the basic public service system and service sector development during the period. Basic elderly care services are part of the basic public service system and the service sector. Therefore, it shall follow the above-mentioned plans. Although the above policies are for the 12th Five-Year Plan, the guiding concepts and basic principles apply to the future development of basic elderly care services.

Development Plan for the Elderly Care Service Industry
The Plan for the Development of China's Undertakings for the Aged During the 12th Five-Year Plan and the Plan for Social Elderly Care Service System Construction (2011–2015) have made detailed deployment on China's elderly care services during the 12th Five-Year Plan period. Beyond the period, China's basic elderly care services should be directed and planned based on the development goals in these plans. The Several Opinions on Strengthening Development of the Elderly Care Service Industry point out, "By 2020, a fully-functional, moderate, and comprehensive elderly care service system covering urban and rural areas, based on families, and supported by communities and institutions, will be established. Elderly care service goods will be more abundant, the market mechanism will be optimized, and the elderly care service sector will develop sustainably and healthily." The above plans provide a direct reference and basis for China's basic elderly care service system from 2020 to 2030.

Ageing Trends
To plan for China's basic elderly care service system and make active response to population aging, we must first scientifically study and judge the basic trend of China's population aging and obtain detailed basic information and data such as the scale, growth rate, age structure, regional distribution, basic needs, and health status of China's elderly population in the future.

Status of Basic Elderly Care Service Development

Outlook or plans for the future development of basic elderly care services must be based on the status quo. Although China's basic elderly care services have developed rapidly, on the whole, (1) the boundaries of the government, the market, society, and individuals are unclear; (2) government power is abused and misplaced; (3) market-oriented development is lagging behind and private capital is insufficient; (4) the contradiction between the supply and demand of basic elderly care services is prominent, (5) the service quality is low and far behind the elderly's demand. China's basic elderly care services must be realistic and establish a mechanism to make up for the shortcomings, so that more and more seniors can access convenient and accessible services.

Lessons from Other Countries

Learning from countries with a highly aging population is an important means to accelerate the development of China's basic elderly care services. These countries' common practices of basic elderly care services include: enhancing the basic elderly care regulation and policy system, increasing market and social guidance and support, strengthening the standardization of basic elderly care services, and vigorously developing and standardizing volunteering.

5.4.2 Development of China's Basic Elderly Care System from 2020 to 2030

Blueprint

The blue print will be centered on solving the contradiction between the supply and demand of basic elderly care services and meeting the elderly's needs for basic elderly care services. The requirements of Several Opinions of the State Council on Strengthening Development of the Elderly Care Service Industry and the Outline of the 13th Five-Year Plan will be met. The overall plan for China's basic elderly care service system from 2020 to 2030 will address the outstanding problems in the system during the 12th Five-Year Plan period. The focus will be on basic elderly care service facility construction, resource allocation, and supply entities. In terms of basic elderly care service facility construction, emphasis will shift from simply "increasing quantity" to both "increasing quantity" and "structural adjustment." In terms of resource allocation for basic elderly care services, emphasis will be placed on the coordinated development of urban and rural areas and regions. In terms of the main provider of basic elderly care services, the decisive role of the market in resource allocation will be fully exerted, emphasis will be placed on promoting the transformation and development of public elderly care institutions, and the government will play its role in policy formulation, supervision, and management.

Overall Requirements
Based on China's conditions, the aim should always be meeting the elderly's growing needs for basic elderly care services. Innovation, coordination, green, opening, and sharing should be upheld in development. Population aging should be practically responded. A basic elderly care service system with balanced service resources, rich content, institutionalized fundraising, specialized workforce, and standardized services should be built, which shall be commensurate with China's economic and social development and demand of population aging.

Balanced Service Resources
The government should play its role in planning, policy guidance, and financial support to promote rational resource allocation and science-based facility layout for basic elderly care services. The development of basic elderly care services in rural areas should be accelerated, and resource allocation for basic elderly care services in urban and rural areas should be coordinated. The pattern and scale of basic elderly care service facilities in urban and rural areas should be reasonably determined according to factors such as the composition and scale of the local elderly population.

Rich Service Content
As China's economy and society develop, social security capacity is improved, and the elderly's needs develop and change, the content of basic elderly care services should be enriched. It should be extended from basic daily care to medical care, rehabilitation care, assistive device installing, spiritual comfort, wealth management, travel and communication, legal counseling, long-term care, and palliative care. All-round services will better meet the various service needs of the elderly.

Institutionalized Fundraising
Policies on the use of elderly care service funds should be introduced. Funding for basic elderly care services should be included in the budget. A regular adjustment mechanism should be established in which the

budget for basic elderly care services grows with economic development. The above measures will institutionalize fund raising and allocation, making them more scientific. In addition, supervision should be strengthened to effectively protect the legitimate rights and interests of investors, builders, operators, and service recipients of basic elderly care service projects.

Professional Service Workforce
The basic elderly care service workforce should be more professional. Job standards and practice specifications should be developed. Occupational education, on-the-job education, and job skill training should be intensified, and relevant disciplines and teaching material systems should be established and improved. Volunteering should be strongly advocated. More effort should be put in volunteer training to institutionalize and standardize the practice.

Standardized Services
Specifications and standards for basic elderly care services should be developed. Standardization of basic elderly care services should be accelerated. Standards for key areas such as service standards, service requirement evaluation, service management, and service facility construction should be developed and introduced to standardize the development of basic elderly care services.

Basic Principles

Insisting on Securing the Basics
The leading role of the government should be brought into full play. China's basic national conditions of the primary stage of socialism should always be the basis, and the principle of "doing our best within our capability" should be insisted. The security should focus on providing basic services for the elderly, such as basic daily care and rehabilitation care. On this basis, with the development of economy and society, the scope should be expanded and the standard should be raised.

Emphasizing Coordinated Development
Home-, community-, and institution-based care services should be developed as a whole. Universal services and personalized services should be combined. The coordinated development of urban and rural areas and regions should be highlighted to promote the equalization of basic elderly care services.

Deepening Institutional Reform

The transformation of government functions should be accelerated, with less administrative intervention and greater policy support and guidance to stimulate the vitality of service providers. Service provision methods should be innovated, and supervision and management should be intensified to improve service quality and efficiency. The management system should be better organized, the responsibilities of departments should be defined, and the operating mechanism should be improved for unified and coordinated actions.

Improving the Market Mechanism

The government's planning, supervision, and service functions should be strengthened to create a market environment that favors equal participation and fair competition. The decisive role of the market in resource allocation should be brought into full play, to allow the private sector become the main actor for the development of basic elderly care services.

Development Goals

The basic elderly care system, which is based on families and supported by communities and institutions, will be improved. The service content will be richer. The management system will be clear and unblocked. The caliber of the workforce will be significantly enhanced. Fundraising methods will be more diverse and funds will be sufficient. The contradiction between supply and demand will be basically resolved.

Home-Based Care Services

In 2020, the home-based care service network will cover all urban communities. Comprehensive community service facilities and centers for basic elderly care services will be established in more than 90% of townships and more than 60% of rural communities. In 2030, the home-based care service network will cover all urban and rural communities.

Community-Based Care Services

In 2020, community home-based care services will be extended to basic daily care, leisure and entertainment, medical care, spiritual comfort, palliative care, and nursing care for the elderly. Urban and rural community elderly care service facilities will be basically developed. By 2030, urban and rural communities will provide comprehensive services including health management, medical care, rehabilitation and nursing, personalized housekeeping, day care, sports and entertainment, psychological comfort, and an around-the-clock hotline. Urban and rural community elderly care service facilities will be developed.

Institution-Based Care Services

There will be more beds in nursing homes. The number of beds in elderly care institutions will account for 4% of the total elderly population in 2020, and 5% in 2030.

Nursing and rehabilitation services will be prioritized. Government investment will focus on building elderly care institutions with functions of long-term medical care, rehabilitation promotion, and palliative care, and encourage private capital to do so. In 2020, each city above the prefecture (city) level will have at least two specialized elderly care institutions. In 2030, all counties (districts) in China will have specialized elderly care institutions.

The Basic Elderly Care Service Management System

In 2020, the urban and rural basic elderly care service network will be improved. The industry management system will be better organized to implement unified industry management for elderly care institutions and facilities. Public elderly care institutions will be fully restructured. Policies and measures will be implemented to encourage private capital to enter elderly care services. Community elderly care service institutions will be basically developed. Significant progress will be made in the senior caregiver workforce. The accessibility of elderly care services will be improved, and the problem of basic elderly care service demand being unmet will be significantly alleviated.

In 2030, an elderly care service system covering the urban and rural elderly population will be basically established. Long-term care services mainly for seniors with disabilities and dementia will be basically popularized. The home-based care service system will be improved. The pattern of diversified elderly care service provision will be formed. Basic public elderly care services in urban and rural areas will be equalized and the elderly's multi-level care needs will be basically met.

System for Basic Elderly Care Service Workforce Construction

In 2020, the number of senior caregivers with professional training and qualifications will basically meet the elderly's care needs. Among them, the ratio of junior, intermediate, and advanced personnel will be about 6:3:1.

In 2030, the number of professionally trained and qualified senior caregivers will fully meet the elderly's care needs. The ratio of personnel at different levels will be further optimized. The quality of elderly care will be further improved.

Fund Raising and Allocation System for Basic Elderly Care Services

In 2020, investment in basic elderly care services will be included in budgets of governments at all levels. Preferential measures to attract the private sector to launch basic elderly care service projects will be improved. A long-term care insurance system will be piloted. Management measures for basic elderly care service funds will be formulated to define all aspects and links involved in the raising and allocation of basic elderly care service funds.

In 2030, the Central and local governments will establish stable fiscal appropriation mechanisms. Government investment in basic elderly care services will grow faster than regular fiscal expenditure, with an increasing share in regular fiscal expenditure. The long-term care insurance system will be fully implemented to set a firmer financial foundation for the elderly to access the services.

Priorities

Coordinating Basic Elderly Care Service Facility Planning and Construction

Construction plans for basic elderly care service facilities should be reasonably determined, and construction standards for basic elderly care service facilities should be strictly observed. Basic elderly care service facility construction should be incorporated into the overall urban planning. Construction plans for basic elderly care service facilities should be scientifically formulated according to the elderly population's size, composition, and needs. Existing public elderly care service facilities should be fully utilized. Basic elderly care service facilities and their supporting service facilities should be coordinated and rationally arranged. The land for basic elderly care service facilities should be clearly allocated. Engineering construction standards and land use standards for basic elderly care service facilities should be strictly observed in accordance with planning requirements. New residential areas should be equipped with basic elderly care service facilities according to planning requirements and construction standards. These facilities should be planned, constructed, accepted, and delivered simultaneously with the residences. If an established residential area does not have basic elderly care service facilities or the existing facilities do not meet the requirements of planning and construction standards, improvements can be made through purchase, replacement, or lease.

Focusing on the Development of Home-Based Care Services

There should be more support and practical preferential policies for home-based care services. The decisive role of the market in resource allocation should be brought into full play to promote and encourage enterprises and social organizations to enter the home-based elderly care service industry. Branded and large-scale home care service chains should be encouraged to provide more diversified and specialized services. The training of home-based elderly care service workers should be intensified to improve the workers' professional skills. The resources of home-based care services should be further integrated, and a home-based care service information platform should be established. The old age allowance, service subsidy, and nursing subsidy systems should be further expanded. Younger seniors may play an active role. There should be more volunteers for home-based care services.

Vigorously Developing Community-Based Care Services

The construction of community-based care service facilities should be strengthened. In cities, community day care centers, nursing homes, public welfare lottery-funded nursing homes, and mutual aid community-based care service centers should be included in the plan for community supporting facilities. The number of elderly care facilities and branches should be increased. In rural areas, under the development of urbanization and the construction of new rural areas, elderly care beds for day care and short-term care should be set up in township nursing homes, which should be transformed into regional social elderly care service centers. Rural communities where conditions permit may explore the construction of a day care service system for the elderly in rural communities according to local conditions, and provide social services such as day care, short-term care, and catering to empty-nesters and other seniors.

Accelerating Elderly Care Institution Reform

Public social-security elderly care institutions should be well functioning. Government-run elderly care institutions must be practical and applicable, and serve as a safe net. They should focus on providing basic support and nursing services for seniors supported by the government, low-income seniors, and seniors with disabilities or semi-disabilities in financial difficulties. Public elderly care institution reform should be piloted. Management measures for public-owned elderly care service facilities operated by private capital should be formulated. Government-run elderly care institutions should, under the principles of government functions separated from institution functions and supervision separated from day-to-day operations, realize private operation through entrusted management, cooperative operation, and other public constructing and private running methods. The private sector should be supported to set up elderly care institutions with the threshold further lowered.

Practically Enhancing the Rural Basic Elderly Care Service System

New models of home-based care services for rural areas should be developed. Regions where conditions permit should be encouraged to use public funds to build concentrated residences for the elderly and supporting service facilities in the construction of new rural areas to provide basic services. Rural community clinics, disability centers, and other infrastructure should be integrated. Abandoned school buildings, kindergartens, offices, and other collectively owned houses should be renovated and staffed to provide basic elderly care services. The reform and innovation of the system and mechanism of rural public elderly care service institutions should be promoted. Rural public elderly care service institutions should give priority to meeting the needs of centralized support for seniors living in dire poverty in rural areas. Township elderly care institutions should be supported to improve the facilities and open to the public. Their services should be enriched, and the service model should be more innovative. They should play a radiating role to become regional elderly care service centers that provide comprehensive services including centralized support, day care, and home-based care services, with higher operational efficiency.

Actively Promoting the Integration of Basic Elderly Care Services with Medical Care

The integrated development of elderly care and medical care at the community level should be promoted. Community basic elderly care service facilities and community medical care and health service institutions should be encouraged and supported to integrate their resources. Models for cooperation between community medical care and health institutions and home- and community-based care services should be developed. The role of community health service institutions in elderly health management, chronic disease prevention, and rehabilitation care should be strengthened. Based on the home- and community-based care service networks, various accessible basic medical care, health, and rehabilitation care services should be provided for the elderly at the community level.

The integrated development of medical institutions and elderly care institutions should continue to be promoted. Elderly care institutions should be supported to set up internal medical institutions, and medical institutions should also be encouraged to set up elderly care institutions. Medical facilities and elderly care service facilities should be coordinated for integrated development. Medical insurance payment should play a better role in guiding the change, with increased payments for medical services provided by medical institutions in elderly care institutions.

Strengthening Personnel Training

The educational resources of higher education institutions, vocational colleges (schools), and vocational training institutions should be guided and integrated for this cause. Training of professionals of elderly service management, medical care, rehabilitation nursing, nutrition, and psychological counseling should be accelerated. Preferential policies should be introduced to encourage graduates of elderly care service-related majors to engage in elderly care services. Vocational training for senior caregivers should be intensified. Workers who attend elderly care vocational training and vocational skill assessments should be subsidized in accordance with applicable regulations. The state's provisions on the technical rating system for elderly care service workers should be strictly implemented. A system of linking the wages of elderly care service workers with professional skills and service years should be established, and workers' pay should be increased. The qualifications and registration assessment policies for medical institutions and welfare institutions should apply to professionals and technicians at elderly care institutions. Private elderly care institutions and elderly care service enterprises enjoy the same treatment as government-run elderly care institutions in technical title assessment, continuing education, and vocational skill training. A talent introduction mechanism for social workers should be established. Through government procurement of services, social work positions should be piloted in the elderly care service industry.

Increasing Public Financial Investment
Financial investment in basic elderly care services should be continuously increased so that the services can practically "secure the basics." The development of basic elderly care services should be made a key area of basic public services and important livelihood projects, and included in the public budget. A financial investment growth mechanism that adapts to population aging and the development requirements of undertakings for the aged should be established. A greater proportion of lottery public welfare funds at all levels should be allocated to develop basic elderly care services. The Central Government should increase its support for rural areas and ethnic minority areas.

Approaches

Improving the Elderly Care Service Management System
The mechanism for undertakings for the aged should be improved, which shall feature the unified leadership of the CPC Committee, government administration according to law, close cooperation between departments, active participation of social organizations, joint measures and concerted efforts. The general pattern of undertakings for the aged should be formed. Departments involved in elderly care affairs should attach great importance to the development of basic elderly care services, and earnestly perform their duties of overall planning, policy support, fund guidance, example-based demonstration, supervision and management. The work coordination mechanism should be further strengthened, and Committees on Ageing at all levels and their offices should play the role of comprehensive coordination, supervision and inspection, and advising and assistance. The development and existing problems of the elderly care service industry should be put under regular analysis, and policies and measures should be developed and introduced to accelerate the development of basic elderly care services. Tasks for the development of the elderly care service industry should be conscientiously completed. Elderly care service resources from all parties should be integrated to form a force of joint management and overall promotion.

Intensifying Reform and Innovation Efforts
Pursuits to emancipate the mind and deepen reform should continue. Efforts should be increased to revolutionize institutional mechanisms, policy systems, development models, inspection, and supervision. There

should be deeper investigation and research, intensified innovation, and effective policies on issues such as the social security system for the elderly, the construction of service workforce, the capital investment mechanism for basic elderly care services, the methods of government procurement of services, the market access and supervision of basic elderly care services, the comprehensive utilization and development of community-based care service resources, and the participation of the private sector in basic elderly care service development.

Enhancing Institutions to Promote Standardized Development

The system for government procurement of basic elderly care services should be enhanced. The recipient of services in government procurement should be clarified and the efficiency of public resource utilization should be improved. A guiding catalogue for government procurement of basic elderly care services should be formulated according to economic and social development, financial affordability, and the elderly's basic needs, which should clearly define the type, nature, and content of services, and be adjusted in due course. The basic elderly care service evaluation system should be enhanced, new evaluation organization models should be discovered, the evaluation index system should be improved, and an evaluation process and supervision mechanism should be established. The basic elderly care service standard system should be enhanced, the basic elderly care service standards should be observed, and the basic elderly care service market should be standardized. Elderly care service dispute settlement measures should be improved to protect the rights and interests of both suppliers and demanders of elderly care services and promote the rapid development of elderly care services. An accidental injury insurance system for the elderly should be established, and a reasonable risk sharing mechanism for the elderly care service industry should be built. A basic elderly care service statistical system should be established, and the basic elderly care service development evaluation and monitoring index system should be improved to comprehensively, accurately, and timely reflect the development of basic elderly care services. Through system building, plan implementation will be guaranteed.

Coordinating the Plan for Basic Elderly Care Service Development

The development of basic elderly care services should be incorporated into the national economic and social development plan and listed as a key development area of the service industry. A basic elderly care service system that is commensurate with the level of economic and social development and aims to meet the elderly's basic elderly care service needs should be established. Medium- and long-term development strategic plans should be formulated and implemented as responses to the challenges of basic elderly care services. The basic goals, basic tasks, development steps, and major measures for the development of basic elderly care services should be set. The development of the elderly care service industry should be supported, which will focus on elderly living care, elderly health services, and livable housing for the elderly. A basic elderly care service evaluation system should be established to promote the equalization of basic elderly care services.

Actively Promoting the Private Sector to be the Main Provider of Basic Elderly Care Services

The government should basically withdraw from the direct production of basic elderly care services. The system and mechanism for government procurement of basic elderly care services should be improved. Preferential and support policies in investment and financing, land supply, tax incentives, and subsidy support for the private sector to invest in the establishment of basic elderly care service institutions should be improved and strictly implemented, so that the private sector becomes the main provider of basic elderly care services.

CHAPTER 6

Modernization of Basic Public Cultural Services

6.1 Understanding Public Culture, Public Cultural Services, and the Modernization of Public Cultural Services

6.1.1 Basics of Public Culture

Culture is necessary for maintaining the common values of the nation and for citizens to exercise their cultural rights. It is essential in maintaining the ideology of the ruling party. In the Chinese context, public culture refers to the sum total of public cultural goods, activities, services, institutions, facilities, and forms to communicate and spread ideology, mainstream social thoughts, basic morality, basic cultural knowledge, and basic sports activities of the masses, which are led by the ruling party and administrative organs and engaging the general public, to meet the basic cultural needs of the public and protect the basic cultural rights and interests of the public.

Public culture mainly includes:

First, public cultural content, such as mainstream ideology, mainstream social thoughts, basic morality, basic cultural knowledge, art and cultural heritage, and basic sports activities of the masses.

Second, public cultural goods and activities, such as media-based public cultural goods, graphic goods in the form of pictures and texts,

K. Dong, N. Wei, *Transforming China's Public Services*,
https://doi.org/10.1007/978-981-99-3943-5_6

three-dimensional goods in the form of audio and video, and virtual goods in the digital form, and performance activities to display public culture.

Third, public cultural institutions and facilities, including institutions that focus on creating and producing public culture, such as publishing houses, theaters, and art troupes; institutions that integrate the creation, production, and communication of public culture, such as radio stations, TV stations, and the emerging online media; institutions and facilities that focus on disseminating public culture and providing public cultural activities (the public cultural service system), such as cultural centers, museums, libraries, art galleries, science and technology museums, archives, gymnasiums, memorial halls, martyrs' cemeteries , workers' cultural palaces, (youth's) children's palaces, historical and cultural parks, and networks of public cultural service facilities.

6.1.2 *Understanding Basic Public Cultural Services*

Basic public cultural services can be understood at two levels. The first level is public cultural services, which refer to the process in which the government provides the above-mentioned public cultural goods and services to the public. In a broad sense, this process includes both the decision-making process and the implementation and feedback processes, both the creation and production, circulation, and provision of public cultural goods, and the consumption and feedback of the public. The second level is basic public cultural services, which refer to the basic, regular, direct-to-all public cultural services. They concern the basic cultural needs and rights of the public. Therefore, the government must be responsible to ensure the universal, convenient, and fair access to such public cultural services for people of all classes in every part of the country. This distinguishes them from general public cultural services.

Basic public cultural services mainly include the following five categories:

First, the basic reading service, or "reading books and newspapers" service. It includes establishing public libraries (stations or rooms) and study rooms, which are open to the public to read books, newspapers, and periodicals for free.

Second, basic radio and TV services, or the "listening to the radio and watching TV and movies" services. It includes establishing public radio stations and TV stations for the public to listen to radio and watch TV for free. It also includes the "culture going to the countryside" service, that is, to build open-air theaters for the public to watch movies for free.

Third, basic cultural and sports activities, where the public can watch plays, operas, performances, and other cultural and artistic performances for free or at a low price, and participate in or carry out mass cultural activities in cultural centers, mass art centers, gymnasiums, (youth's) children's palaces, and cultural palaces.

Fourth, basic public cultural appreciation activities, where the public can watch and appreciate various public cultural goods in museums, exhibition halls, memorial halls, former residences of celebrities, revolutionary sites, natural heritage parks, and cultural heritage parks for free or at a low price. They are also known as cultural tourism services.

Fifth, basic information services, where the public can browse public cultural information for free or at a low price.

6.1.3 Understanding the Modernization of Public Cultural Services

The Decision of the CPC Central Committee on Major Issues Concerning Comprehensively Deepening Reforms adopted at the Third Plenary Session of the 18th CPC Central Committee envisioned "a modern public cultural service system." The core of the modernization of basic public cultural services lies in the "modernization of public culture" and the "modernization of services." The former refers to the content of public culture. It means that public culture must keep pace with the times and present modernization features that are compatible with the economic and social modernization. Its basic requirements are to continuously develop advanced culture, innovate traditional culture, insist on supporting and guiding popular culture, effectively filling cultural lags, and resolutely resist harmful culture. The latter refers to the requirements for public cultural service provision. It means that the services, including its concept, content, means, and mechanism, must be modernized. Its basic requirements are to cover urban and rural areas, be convenient and efficient, meet basic needs, and promote fairness.

The modernization of service content means adding new public cultural content so that some widely used new knowledge, widely accepted new concepts, new arts, and mass activities become new public cultural content. In 2030, China will basically complete its industrialization and enter a new era dominated by the service industry. New public cultural content will be part of basic public cultural services along with the modernization process. For example, the popularization and application of digital knowledge, public square dance training, provision of some public information, and free digital TV channels and digital movies are likely to be included in basic public cultural services.

The modernization of service means refers that service methods and means are armed with new technology along with the advancement of science and technology, making basic public cultural services more personalized and the provision more efficient. Importantly, in 2030, Internet technology and other high-tech technologies will be applied deeply and widely. As a result, basic public cultural services must make extensive use of these modern technologies to match with the technical means used by the public every day. For example, most public cultural goods and services will be synchronized on the Internet, and people can obtain information and enjoy related services from the Internet anytime, anywhere. Therefore, providers of public cultural goods and services will meet the needs and consumption of the public mainly through modern information technology.

The modernization of service mechanisms refers that service policies and institutional mechanisms are modernized along with the modernization of national governance concepts and administrative management systems. They will be more democratic, transparent, and equal, and better demonstrate fairness and justice. For example, in 2030, public cultural service policies can adopt the opinions and suggestions of stakeholders, especially local residents, in the aspects of formulation, implementation, and feedback, so as to maximize democratic decision-making. Another example is that in 2030, all kinds of cultural and sports facilities and service resources can be effectively integrated, making the best use of material and human resources.

Legislation of service policies. As the Resolution of the Central Committee of the Communist Party of China on Major Issues Concerning the Enhancement of all-round Law-based Governance was deliberated and adopted at the Fourth Plenary Session of the 18th CPC Central Committee, the concept of law-based governance has gradually taken root in the hearts of the people. Institutions and systems of law-based governance have been improved. More public cultural service policies have become law. Various service policies have been basically incorporated into the legal system and law-based governance. Random changes and alteration of public cultural policies have been significantly eliminated. Policies must be adjusted in accordance with legal procedures.

6.2 History and Current Status of and Problems in Basic Public Cultural Services

6.2.1 History of Basic Public Cultural Services

From the Founding of the People's Republic of China and the Beginning of Reform and Opening Up

(1) Changes in the public cultural administration system. Just after the founding of the People's Republic of China, the authorities in charge of public culture mainly included the Publicity Department of the CPC, the Ministry of Culture, the General Administration of Information, and the General Administration of Publication. After the General Administration of Information was abolished in 1952, its subordinate Broadcasting Board and Xinhua News Agency were placed under the Culture and Education Commission, and the General Administration of Publication took over newspaper administration. After the General Administration of Publication was abolished in 1954, newspaper and publication administration were placed under the Ministry of Culture. In 1954, the Culture and Education Commission was abolished, and the competent authorities for public culture became the Ministry of Culture and the Broadcasting Board. The Ministry of Culture was abolished in 1970 and reinstated in 1975.

(2) Development of the public cultural service system. In 1949, there were only 55 libraries, 896 cultural centers (including mass art centers), and 21 museums in China.[1] In 1952, there were 98 radio stations nationwide, with 65 domestic radio programs. In 1958, Beijing TV was established as the first TV station in the People's Republic of China.[2] According to the survey and statistics of the National Press Conference in the spring of 1950, there were 253 public and private newspapers in China. By the end of 1956, there were 101 publishing houses in China.[3] In 1961, the State Council announced the first batch of 180 major historical and cultural sites protected at the national level. In 1982, the State Council approved the first batch of 24 national historical and cultural cities. In 1987, China successfully had six sites inscribed as world heritage for the first time.

[1] http://www.china.com.cn/news/200909/14/content_18522783.htm.

[2] Fang Hanqi, *General History of Chinese Journalism Vol. 3* (Beijing: China Renmin University Press, 2004): 31.

[3] Cheng Zhen, "The Chinese government's publications and government information dissemination," *Journal of The National Library of China*, no. 4 (2001): 23–29.

Before Deepening the Cultural System Reform

(1) Changes in the public cultural administration system. During that period, the main organs in charge of public culture were the Publicity Department of the CPC, the Ministry of Culture, the Ministry of Radio and Television, and the Press and Publication Administration. In 1982, the former Ministry of Culture, the Commission for Cultural Relations with Foreign Countries, the State Publication Administration, the State Administration for Protection of Cultural Relics, and China Foreign Language Publication Administration merged into the new Ministry of Culture. In the same year, the Ministry of Radio and Television was established based on the former Central Broadcasting Board. In 1986, the Ministry of Radio and Television and the Bureau of Culture, Film, and Television merged into the Ministry of Radio, Film, and Television. In 1985, the Ministry of Culture established the National Copyright Administration, changing the former Publication Bureau of the Ministry of Culture to the National Publication Bureau. The National Publication Bureau and the National Copyright Administration were "one institution with two names." In 1987, the National Publication Bureau was abolished, and the National Press and Publication Administration was established, which was directly under the State Council. At the same time, the National Copyright Administration was retained and the form of "one institution with two names" continued. In 1998, the Ministry of Radio, Film, and Television was changed to the State Administration of Radio, Film, and Television. In 2001, the Press and Publication Administration changed its name to the General Administration of Press and Publication and was upgraded to a ministerial-level agency.

(2) Development of the public cultural service system. By the end of 2011, the Cultural and Information Resource Sharing Project covered 100% of provinces (autonomous regions and municipalities directly under the Central Government), 99% of districts and counties, 83% of townships, and 99% of administrative villages. In some provinces (autonomous regions and municipalities directly under the Central Government), village-level coverage was extended to natural villages.[4] There were 2952 independent public libraries at or above the county level, 3285 cultural centers (including mass art centers), and 40,390 township (sub-district) cultural stations. The goal that "each county has at least a library and a cultural center and a township has at least a comprehensive cultural

[4] http://www.ndcnc.gov.cn/fagui/fagui/201302/t20130228_582063.htm.

station" was basically achieved. A six-level (state, province, municipality, country, township, and village) network of public cultural facilities was basically established.[5] By the end of March 2012, 94% of the projects subsidized by the central funds under the National Township Comprehensive Cultural Station Construction Project were completed. Among them, the completion rate was over 99% in 11 provinces, and provinces (autonomous regions and municipalities directly under the Central Government) such as Fujian, Anhui, Hubei, Tibet, and Chongqing completed their construction tasks.[6] After the implementation of the Rural Movie Screening Project, 8.12 million movies were screened in rural areas across China for 1.8 billion moviegoers in 2011. The goal of "one movie in one village every month" was basically achieved.[7] The Extending Radio and TV Coverage to Villages Project covered all administrative villages and natural villages with more than 20 households and connected to the power grid. More than 600,000 rural reading rooms were built, covering all qualified administrative villages across China. The planned tasks were realized three years ahead of schedule.[8] By the end of 2011, there were more than 200 million cable network users nationwide. All central and provincial radio and TV stations basically achieved digitalization and networking. Digital terrestrial TV covered 337 large and medium-sized cities across the country, with a digital penetration rate of about 57%, which was less than 1% in 2004. There were 36 agricultural radio frequencies, 13 agricultural TV channels, 33 children's channels, 5 satellite animation channels, and 4 documentaries channels nationwide.[9]

6.2.2 *Current Status of Basic Public Cultural Services*

Up to now, China has basically built a public cultural service system covering urban and rural areas based on five cultural projects that benefit the people, including the Rural Movie Screening Project and the Rural Reading Room Construction Project. The basic cultural rights and interests of the people are better guaranteed. China is the world's third largest

[5] http://www.gov.cn/gzdt/201206/20/content_2166028.htm.

[6] http://www.mcprc.gov.cn/sjzz/cws_4693/whtj_cws/tjgb/201308/t20130823_387282.htm.

[7] Cai Fuchao, "Let culture reach everyone: The development of China's radio, movie, and TV public service system," *Qiushi*, no. 17 (2012): 22–24.

[8] https://www.xinhuanet.com/18cpcnc/zhibo/20121111a/wz.htm.

[9] https://www.xinhuanet.com/18cpcnc/zhibo/20121111a/wz.htm.

movie producer and largest TV show producer by annual output. In 2014, the National Reading Campaign attracted more than 800 million participants. By the end of 2014, 3117 public libraries at or above the county level, 3313 cultural centers (including mass art centers), 41,110 township (sub-district) cultural stations, 3473 public museums and memorial halls, more than 1300 workers' cultural palaces, more than 700 youth's and children's palaces, more than 350 science and technology museums, and more than 3000 youth extra-curricular activity centers were built across China. 2780 public museums, 347 patriotic education demonstration bases, and 43,510 public libraries, art galleries, and cultural centers (stations) across the country were opened free of charge. By the end of 2014, there were 6337 registered cultural volunteer service organizations at or above the county level, with 910,000 cultural volunteers and 1.74 million social sports instructors. A primary-level cultural workforce that comprises of full-time and part-time workers took shape.[10]

By the end of 2014, 2564 radio and TV broadcasting agencies and more than 21,000 transmission stations were established nationwide. There were more than 120,000 publication distribution entities and more than 170,000 distribution outlets. More than 72,000 newspaper reading boards and screens were constructed. More than 420,000 rural sports and fitness projects were completed. Almost all rural areas connected to the power grid in the country had access to radio and television, with the comprehensive population coverage of radio and TV programs of 97.99% and 98.6%, respectively. The National Cultural and Information Resource Sharing Project built 35,500 township (sub-district) primary-level service stations and 700,000 village (community) primary-level service stations, basically covering all townships and administrative villages. The Rural Movie Screening Project built 252 digital cinemas, about 50,000 screening teams, played 8 million movies for about 1.5 billion audience every year. There were more than 7200 digital screens in county-level cities across China. In 10 provinces (autonomous regions and municipalities directly under the Central Government), all county towns had digital cinemas. The Rural Reading Room Project built 601,000 reading rooms and 16,000 satellite digital rural reading rooms for remote areas. 69 high-definition TV channels, 29 Internet radio and TV stations, and 605 Internet audio-visual program service agencies were open. The number of mobile multimedia radio and TV users reached 45 million. The digital and

[10] http://www.npc.gov.cn/npc/xinwen/201504/23/content_1934246.htm.

bidirectional transformation of the cable TV network was promoted. There were nearly 190 million digital TV users nationwide. In 2013, 100,000 hours and 40,000 hours of ethnic minority radio and TV programs were produced and dubbed, respectively. In 10 border provinces, 810 "Border Digital Culture Corridor" township service stations and 3104 digital cultural stations were built.[11]

By June 22, 2014, China had 47 world heritage sites, ranking second in the world. Thirty-eight intangible cultural heritage elements in China were inscribed in the UNESCO Intangible Heritage List, ranking first in the world. The country had more than 4295 major historical and cultural sites protected at the national level and 126 cities in China was remarked as Historical and Culture City, which is an honour given by Ministry of Culture and Tourism of PRC. A total of 1372 representative elements of national intangible cultural heritage in four batches and 1986 national representative successors were examined and recognized. One hundred demonstration bases for productive preservation of intangible cultural heritage in two batches were named, and 18 national experimental zones for cultural ecology preservation were established.[12] By December 2015, the number of Chinese Internet users reached 688 million, and the Internet penetration rate reached 50.3%. Among them, rural users accounted for 28.4%, an increase of 9.5%. At the same time, the number of mobile Internet users reached 620 million, and 90.1% of them accessed the Internet through mobile phones.[13] The Internet became a new way and vehicle of cultural life in China.

6.2.3 Major Existing Problems

Emphasizing the Form over the Content

In the development and construction of public culture, the form is overemphasized, while the content, which should be the core and essence, lacks innovation. For example, healthy-themed songs, dances, plays, TV shows, and movies focus on lighting, stage art, and 3D effects to gain attention. They are sung and performed over and over again, but there are few new public cultural works that can get to the bottom of people's hearts. On the other hand, too much effort is put into building vehicles of public culture. Two types of vehicles are referred to here, namely hardware

[11] http://www.npc.gov.cn/npc/xinwen/201504/23/content_1934246.htm.

[12] http://www.npc.gov.cn/npc/xinwen/201504/23/content_1934246.htm.

[13] http://media.people.com.cn/n1/2016/0123/c4060628078214.html.

and software. The former is infrastructure, which will be discussed in detail later. The latter is the direct form of public cultural content. For example, much effort is devoted to presenting certain public cultural content in various forms, such as books, audios, videos, the Internet, audio-visual products, opera, and dancing, but the content is neglected.

Emphasizing Supply over Demand

First, some public cultural goods are produced and supplied according to "instructions" or "plans" with little regard for the needs of the public. Second, a considerable number of public cultural goods in society are supplied based on "captain's call" without understanding demand in social practice and grassroots. For example, the supply of a large number of public cultural infrastructure and the communication and popularization of high public cultural goods are likely to fall into the power supply chain of the local government to create a "cultural highland" and seek "political achievements."

Emphasizing Independent Operation over Overall Coordination

In the complicated bureaucratic organizational system, many systems and provinces, municipalities, counties, and even townships have their own cultural centers (stations) or public cultural communication and education institutions, resulting in serious duplication of construction. For example, many central ministries and commissions and provincial departments and bureaus have their own communication and education centers; almost every county has a cultural center and library; and every township has a comprehensive cultural station. As a result, the public cultural goods provided by the communication and education centers of provincial departments and bureaus are likely to overlap the products of the communication and education centers of ministries and commissions at higher levels. Another example: The township comprehensive cultural stations in county towns are unlikely to attract many visitors, as the public is more willing to go to county-level cultural centers and libraries for cultural activities.

Emphasizing Cities over Rural Areas

First, the expenditure on cultural undertakings in rural areas is much lower than that in cities. According to statistics, during the 11th Five-Year Plan period, the expenditure on cultural undertakings in rural areas accounted for less than 30% of the national total, about 40% lower than that in cities. However, the rural population is apparently larger than the urban population even by now. Second, the encroachment of public cultural facilities in rural areas is far more frequent than in cities. Many rural reading rooms are built in CPC village committees and are usually used for meetings. Moreover, many rural public cultural facilities are rented out for other purposes, and the phenomenon of diversion is also prominent.

Emphasizing Facility Construction over Activities, Management, and Utilization

First, activities are neglected. It is a prominent problem that the construction of public cultural facilities is in full swing but the motivation to carry out public cultural activities is insufficient. On the whole, most of the public cultural activities are "forced" to be carried out because they are an important part of the appraisal, instead of being carried out actively by public cultural activity providers.

Second, daily management services are neglected. It is a prominent problem that daily management is loose and services are not good. Some public cultural facilities have a management system. However, in practice, it is not fully implemented due to the extremely low utilization rate of the public. In addition, public cultural facilities age rapidly due to a lack of routine maintenance and timely update, and become some "white elephants."

Third, the overall utilization is insufficient. Except for a few public cultural facilities in first-tier cities and some second-tier cities, such as the National Library of China, the National Museum of China, and other popular cultural centers (stations), a large number of public cultural facilities are rarely visited. Many cultural centers are used for other purposes. Some cultural rooms and libraries are left unattended. Some sub-district (township) cultural stations are built in the office buildings of sub-district offices. People are not allowed to enter or carry out activities for the concern of affecting the work of the offices. Take Beijing as an example. In 2012, the Beijing Academy of Social Sciences carried out a special investigation on the "cultural boom and prosperity in the capital." The survey results showed that 90% of the respondents said that they visited public cultural venues, such as cultural centers, archives, primary-level libraries, and cultural stations, no more than five times a year. More than half of the citizens never went to those cultural venues during the year. It indicated a severe situation of resource idle.[14]

Emphasizing Large Public Cultural Service Facilities over Facilities in Primary-Level Communities and Rural Areas and Small Facilities
One reason is the political-achievement orientation. Another reason is that most public cultural service facilities in primary-level communities and rural areas are small, while their construction must comply with various procedures and standards stipulated in government policies. Therefore, compared with large facilities, their "price-performance ratio" is not high, and the driving force is insufficient.

Emphasizing General Population over Special Groups
It is a prominent problem that although there are public cultural facilities for special groups such as the elderly, minors, women, and migrant workers, there are very few activities that are in normal operation and designed for them. The main reasons are: First, financial constraints. If the needs of special groups are specifically considered and tailored public cultural services are created for them, it will cost much more than for the general population. Second, special groups are vulnerable. It is not easy for them to form a joint force to put pressure on policy makers of public cultural service facilities. As a result, they are less likely to be considered in policy makers' agendas.

6.3 Outlook for the Modernization of China's Basic Public Cultural Services in 2030

6.3.1 *Analysis of Public Cultural Needs Before 2030*

Content of Public Culture

Needs of the State for Public Cultural Content
It is mainly the public cultural content of the ruling party and government.

First, socialist culture will become an important part of the ruling party's needs. In 2030, the leadership of the CPC in China's political, economic, and cultural life will be strengthened like never before. Its

[14] http://www.wenming.cn/whhm_pd/yw_whhm/201301/t20130121_1038336.shtml.

ruling recognition will be greatly improved as it leads the country to successful modernization, basically achieves common prosperity, and strictly combats corruption. At that time, the CPC needs to take this opportunity to firmly establish a people-centered approach in ideological work and deeply study the characteristics of the cultural needs of the people in the new era to ingeniously and vigorously promote socialist culture with Marxism, socialism, and the CPC as the core. It includes public cultural goods introducing the history of the CPC; praising CPC leaders; introducing the CPC's basic theory, line, principles, and policies; introducing the achievements made by the country and the people in various fields of construction under the leadership of the CPC; and praising the people's dominant position. In this way, the CPC will consolidate its dominance in the ideological sphere and change the current status quo in which socialist cultural communication is not "down to earth" and tends to be rigid and dogmatic.

Second, the core socialist value system will become the core of the ruling party's needs. With the great achievements in the construction of socialism with Chinese characteristics and the vigorous promotion of the core socialist value system, the system will become more popular among the people, and a stable value system will form.

Third, the cultural goods and services of the state apparatus will become an important part of the needs of governments at all levels. The government will transform into a "service provider" for society and the market. To this end, self-promotion will become one of the important needs of government departments at all levels. Specifically in public cultural goods and services, the government and government departments will need more public cultural goods that introduce the organization, functions, policy documents, handling procedures, and contact information of public servants of government departments at all levels.

Needs of Society and the Public for Public Cultural Content

First, the important position of traditional culture in public cultural content will be further highlighted and effectively innovated. Justin Yifu Lin believed that by 2030, traditional Chinese culture with "benevolence" as its core ethical value may be revived along with the development of China's economy.[15] We also believe that traditional Chinese culture will see a real revival and innovation by then. First, official institutions will vigorously

[15] www.bjnews.com.cn/news/2011/11/05/162220.html.

advocate and promote traditional culture. The publicity departments of CPC committees at all levels, as well as the cultural, education, radio, movie and television, and even tourism departments of governments at all levels will introduce various incentives, such as incorporating traditional culture into the national education system. Second, non-governmental institutions will invest considerable human, financial, and material resources to develop traditional culture. Traditional culture will be revived in society. Third, through modernization, citizens will integrate modernity and tradition more calmly and reasonably. Chinese people will be more diverse with more distinguished oriental characteristics.

Second, the proportion of mass culture in the public cultural content system will be greatly increased, and it will truly be deeply rooted among and welcomed by the people. Initiatives to improve public etiquette and ethical standards that are being carried out extensively, such as Model Cities, Model Villages and Towns, Model Entities, Model Families, and Model Schools, "Learning from Lei Feng" volunteer service activities, the National Reading Campaign, as well as selection activities such as Good Family Values, Good Local Values, Good School Values, and Good Profession Values, will be truly "down to earth" and be warmly welcomed. All walks of life, localities, schools, entities, and the general public will flock in these activities. The social ethos will further improve. Government-oriented mass cultural activities will come in a variety.

Third, leisure culture will be incorporated into public cultural content for the first time. Leisure culture mainly refers to general leisure and entertainment, that is, people's daily cultural leisure. It will become part of the important content of the public cultural service system in 2030, which includes traveling, board games, card games, watching performances, and physical fitness. The government will provide some basic cultural and leisure services in a centralized manner. In addition, the government will increase the effective guidance of popular culture and the crackdown of harmful culture to further purify the cultural environment. For example, the government will encourage and publicize popular artists who pursue both professional excellence and moral integrity, and punish cultural vices.

Trends of Cultural Needs

First, the substantial increase in the demand for public culture is mainly manifested in two aspects:

On one hand, the demand for the existing basic public culture will continue to increase, and the content of the demand will be more diverse. For example, book and newspaper readers will no longer be satisfied with having books and newspapers to read but will require the government to provide such services in a more timely, richer, and more targeted manner. Another example: The Culture Going to the Countryside campaign will further require holding local activities regularly, and there will be higher requirements for the quality of cultural services going to the countryside.

On the other hand, some of the current special cultural needs will be regarded as public cultural needs. With the continuous economic growth, some of the special cultural needs currently produced and provided in the market will be regarded public cultural needs, and all or most of them will be provided by the government. For example, with the rapid development and advancement of Internet technology, Internet infrastructure will be provided mainly by the government. Citizens will pay a much lower price for Internet access.

Second, public cultural needs will be inclined to integrate with needs in other fields. It will be easy for them to form a comprehensive demand system. Any effect on economic, political, and cultural needs will affect the overall satisfaction with the government. Specifically, the weight of public cultural needs and other social welfare needs will be greatly increased, and whether they are met or not will affect the national happiness index, which in turn affects the satisfaction with the government.

Third, public cultural needs will be met in a timelier manner. For example, public cultural goods will be digitized and synchronized online. Public cultural services will be booked and enjoyed online, or be available within the village and the community. In other words, meeting public cultural needs requires very little time or little distance.

Finally, local public cultural needs will be further highlighted. By 2030, in addition to the national basic public cultural needs, local public cultural needs within their respective administrative jurisdictions will be greatly highlighted. In economically developed areas, it will be likely that there are local needs from the provincial level to the township level, while underdeveloped areas may have fewer local needs for public culture.

6.3.2 Basic Public Cultural Needs in 2030

Major Service Providers

A reasonable pattern in which the government, society, and the market together supply public culture and provide public cultural services, and jointly contribute and share will basically take shape. Among them, the government plays a fundamental and leading role, and the society and the market are auxiliary and supplementary players. The government mainly plays a role in the political and basic areas. Political areas mainly include memorial halls for revolutionary martyrs, former residences, revolutionary sites, and cemeteries for revolutionary martyrs. Basic areas mainly include basic public cultural facilities and their support goods and services, such as libraries and the books and periodicals in them, science and technology museums, and the scientific and technological models in them. Society and the market play a role mainly in other areas, which are not closely related to politics or universal, including featured museums and exhibitions.

Service Provision Methods

Means of Government Provision

The main method is to purchase public services with fiscal expenditure. It means to obtain the production rights of public cultural infrastructure, including the hardware facilities and basic software equipment of public cultural venues, by calling for bids in the market and awarding bids to enterprises. Other methods include government subsidies to the supply side or demand side and direct government provision.

Means of Society and Market Provision

The main method is non-profit organizations (NPOs), enterprise organizations, and individuals providing public cultural services to society free of charge.

First, provision by NPOs. According to the forecast of the Ministry of Civil Affairs, there will be more than 800,000 NPOs in China in 2020 and 1.693 million social organizations registered nationwide in 2030.[16] At that time, the external environment will be more relaxed, and the government and enterprises will increase their funds to a record high level. Therefore, NPOs will be much more active in social welfare undertakings, especially public cultural undertakings. In addition, as population aging in China will accelerate in 2030, there will be a large number of healthy elderly

[16] The Ministry of Civil Affairs of the People's Republic of China, *China Civil Affairs' Statistical Yearbook 2011* (Beijing: China Statistics Press, 2011): 112.

people. In cities, they will become the main force of NPOs engaged in public cultural services. In rural areas, most of them will join elderly associations along with the revival of traditional Chinese culture. Those associations will become the main vehicle to maintain and promote traditional culture and support the development of new public culture in rural areas.

Second, provision by enterprise organizations. With the full establishment of the socialist market economic system, enterprises will be more capable of providing public services within their capabilities and expertise. Moreover, the provision of public services will become an important part of their core competitiveness. Most enterprises will see it as an important factor to increase potential customers, stimulate potential demand, or an extension of after-sales service. Driven by these two factors, more and more enterprise organizations, especially cultural enterprise organizations, will provide public culture for free. For example, textbook publishing houses will host exhibitions of the latest edition of textbooks for primary and secondary schools, which will be free for primary and secondary school students to visit.

Third, provision by individuals. In 2030, there will be a large number of wealthy individuals enthusiastic about social welfare in China. In fact, this trend has already emerged. For example, in coastal areas such as Jiangsu and Zhejiang and big cities such as Beijing, Shanghai, Guangzhou, and Shenzhen, there have been citizens who run their own museums and open them to the society for free. By 2030, this practice will be very common, especially in the culture field. There will be a considerable number free cultural exhibitions held by cultural celebrities, such as free calligraphy and painting exhibitions by calligraphers and painters and free concerts by singers.

Service Provision Standards

In 2030, all kinds of public cultural services will be put under the government's unified standard framework, that is, the government will formulate unified service delivery standards. It will include basic requirements for the content (compliance with law and socialist ethics), hardware (the compliance of infrastructure with technical standards), security (guarding the personal safety and privacy of users and visitors of public culture) and procedures (the approval system or filing system).

Service Coverage

Covering All

Covering all means the inclusiveness of basic public cultural services. Regardless of region, ethnicity, gender, age, education level, occupation, class, or other background factors, all people can enjoy basic public cultural services. A national protection network for basic public cultural rights will be formed. In other words, in the access to basic public cultural services, the existing significant regional differences, urban-rural differences, and differences between the rich and the poor will be greatly narrowed.

Covering Organizations at All Levels

The Central Government, provinces, municipalities, districts (counties), townships and sub-districts, villages and communities will be covered by basic public cultural services, forming a vertically consistent basic public cultural service system. In other words, the work focus will effectively go down. The basic public cultural service system in primary-level communities and rural areas will be developed and improved. The basic public cultural service system will be extended to the most primary-level urban and rural communities.

More Equal

After years of coordinated development in terms of category, quantity, and quality, the basic public cultural services for all will be more equal between urban and rural areas and between regions, and the equity will be guaranteed. That will be an important indicator that the fiscal transfer payment is continuously intensified and the fruits of reform and development are shared by the people across the country.

More Accessible

Citizens will be able to enjoy basic public cultural services more conveniently and efficiently. Each administrative village and community will have a brochure to introduce local and surrounding public cultural service facilities. In terms of specific service content, some service content can be enjoyed at home, such as listening to radio and watching TV on the Internet. Most services can be enjoyed in the community or village. Public cultural service stations for free reading, watching movies, watching performances, and physical fitness, will be distributed more intensively and evenly. This is another inevitable result of the continuous economic and social development.

Specific Service Content

In 2030, China's basic public cultural services will fundamentally get over the unbalance between supply and demand. The public cultural supply will basically match the demand of the country and society. This is mainly manifested in:

Press and Publication

First, public cultural goods on the ruling party will successfully take root in society and be basically supplied for free. These products include CPC newspapers and periodicals, public documents of the CPC, and the CPC's communication materials. They will be made in a popular discourse system and "down to earth," and be warmly received by the public. At the same time, they will be basically distributed free of charge on demand, fully cover the whole Party, and be equally distributed in urban and rural areas. In public places with a large flow of people, such as docks, squares, hotels, airports, railway stations, and ferries, there shall be some CPC newspapers and periodicals for people to read.

Second, the public information goods of governments at all levels will be timely and comprehensive, and effectively meet the needs of the public. They will be basically available for free in the jurisdiction. The service provision stations will be generally set up in the offices of administrative agencies. They include the list of powers and responsibilities, the procedures for handling affairs, the list of service personnel and their office telephone numbers, government bulletins, departmental regulations, and normative documents of the local government and government departments. These materials will be distributed free of charge to stakeholders in government offices, making it easier for stakeholders to obtain public administration information they need.

Third, news information, newspaper, and periodicals on basic livelihood will be authoritative and accurate, and can basically be provided free of charge to specific groups of people across the country. They refer to books, periodicals, images, and newspapers for general popularization in basic agriculture, education, science and technology, health, culture, and information. For example, basic textbooks will be provided free of charge to primary and secondary school students across the country, funded by the education authorities. Local basic information, including books on historical geography and local customs, travel guides, and basic public service guides, will be provided free of charge in the local area, funded of the civil affairs, tourism, cultural, and other competent departments. All kinds of agricultural planting and breeding techniques are provided free of charge to all rural residents (in households) within the jurisdiction, funded by the agricultural authorities.

Fourth, the basic life information will be timely updated with key points highlighted. It will be distributed free of charge to communities and administrative villages. It includes the contact information, opening hours, charges, and conditions for enjoying basic public cultural services of local education, science, culture, health, and other social service organizations. They will be made into pamphlets and placed in CPC village and community committees or community service centers for free supply. They will be updated in time so that the public can obtain accurate public life information at any time.

Fifth, some publications in ethnic minority languages will be provided free of charge. In addition to the above-mentioned public cultural goods and services, some publications with obvious publicness, such as university textbooks, will also be provided free of charge in ethnic minority areas.

Radio and TV

The main manifestation is that radio and TV programs will be richer in content, classified in more detailed categories, and operated in a more diverse mechanism. At the same time, their publicness will be more prominent.

First, the classification of public radio and TV programs will be further divided and enriched. There will be more non-profit radio and TV programs, and the classification will be finer.

Second, a considerable part of the currently paid commercial radio and TV channels will become free public channels.

Third, there will be more and more movies featuring public culture. With the increasing importance and urgency of integrating social consensus and condensing social mainstream values, there will be more and more government-led, enterprise-produced, and free-supplied movies featuring public culture, especially non-profit micro-movies, which will be popular on various platforms (TV, Internet, etc.).

Fourth, free movie screenings. The sheer volume of TV content makes it hard to choose. By contrast, free screenings of excellent movies will become a new public cultural landscape in 2030. This means that movies that have obvious public cultural characteristics and have been offline will be screened in cinemas for free by the government. In addition, communication and popular science high-definition movies and documentaries introducing China and other countries, natural scenery, human landscape, culture and art, and culture excellence will also be screened by the government free of charge.

Fifth, the translation of radio, movie, and TV programs in ethnic minority languages. With the continuous economic and social development of ethnic minority areas and the constant increase in financial transfer payments, more and more excellent radio, movie, and TV programs will be translated into ethnic minority languages funded by the government. Residents in ethnic minority areas will have access to more radio and TV programs in their own languages.

Art and Culture

First, culture and museology, that is, cultural activities taking place in public cultural infrastructure at all levels, such as cultural centers (stations and rooms), libraries (reading rooms), mass art centers, museums, art galleries, science and technology museums, archives, memorial halls, martyrs' cemeteries, workers' cultural palaces, and youth and children's palaces. All basic services in this area will be available free of charge in 2030.

Second, performance, that is, music, dance, acrobatics, plays, and other theatrical performances. In 2030, the content theatrical performances will place equal emphasis on socialist culture and traditional Chinese culture excellence, and will be presented in a way loved by the masses. Free public performances will be held regularly to meet the growing public cultural and entertainment needs.

Third, conventions and exhibitions, including exhibitions of static artistic works such as art, calligraphy and painting, handicrafts, and stamps. In 2030, the exhibits and forms of the above-mentioned artistic work exhibitions will be based on the cultural characteristics of the exhibition locations, so as to effectively appeal to the local public. There will be a free exhibition in every sub-district and county every quarter to meet the public's growing need for cultural appreciation.

Fourth, mass cultural and artistic activities. They include knowledge dissemination lectures, such as lectures on health knowledge and breeding technology, and collective cultural and artistic activities, such as square dancing, aerobics, and calligraphy, painting, and singing contests. In

2030, each administrative village or community will carry out at least one of the above two types of activities from time to time. If the conditions permit, they can be carried out on a regular basis. Leaflets will be given off and prizes will be provided to attract community residents and villagers to participate in the activities.

Fifth, ethnic minority literature and art. In order to protect the ecological diversity of ethnic cultures and further unite ethnic minorities, the government will provide more support for ethnic minority literature and art, especially the content that is at risk of being lost. Financial support and content preservation from local governments and even higher-level governments will be intensified. Relevant performances and communication will be more frequent and active.

Mass Sports

In terms of mass sports activities, in 2030, communities and administrative villages will be the most basic public activity spaces. In developed areas, each administrative village or community will have a standard physical exercise area, a sports service NPO, and a full-time (part-time) sports instructor. The community or administrative village will regularly hold sports meetings or special events every year. In less developed areas, each administrative village or community will have a standard physical exercise area; each township or sub-district will have a sports service NPO; a full-time sports instructor will make rounds to guide physical fitness activities in administrative villages and communities. The community or administrative village will hold sports meetings or special events from time to time.

In daily life, the National Fitness Program is fully implemented. Radio calisthenics will be promoted and persisted among students and office workers, and also promoted to a certain extent in rural areas.

Internet and Information

First, Internet access services. The charges for Internet access for emerging terminal electronic products such as mobile phones and tablet computers will be greatly reduced. Low threshold and high network speed will be realized.

Second, there will free Internet services in public places. In 2030, people can access to the Internet in public places without passwords and browse the content and relevant information of public culture and other public products for free. The accessibility of basic public services in public places will be realized.

Third, all other forms of public cultural goods will be available on the Internet, such as e-books, digital libraries, digital museums, digital archives, digital parks, videos, and audios. Some of them will be free, while some will require certain qualifications and fees. For example, digital archives are costly in the process of digitization and must meet confidentiality requirements, which needs consumers to pay a fee to reduce the financial burden. Nonetheless, most public cultural goods should be provided free of charge, especially the text, video, and audio information of the CPC and the government that can be made public, as well as healthy-themed songs, movies, and TV series.

Tourism and Leisure

They mainly include natural heritage parks, cultural heritage parks, protected historical and cultural sites, historical and cultural cities, and historical and cultural towns (villages).

First, they will be open to key groups free of charge, mainly including minors, the elderly, people with disabilities, active servicemen, and low-income groups.

Second, the overall ticket fee will be significantly reduced. Their income will mainly rely on differentiated goods and services within the premise, such as specialty goods, catering, and other goods and services.

Third, more and more parks and historical and cultural cities (villages and towns) will be opened for free. The practice of the West Lake in Hangzhou has been extended to the whole country, forming a benign demonstration effect. Tourism culture will be a public cultural brand rather than a source of economic income, and parks will be real public leisure places. In 2030, free parks and historical and cultural cities (villages and towns) will account for more than 50% of the total.

Infrastructure

In 2030, public cultural spaces will be put into use at an accelerated rate, most of which will be regulated, institutionalized, and operated according to detailed rules. The current situation of emphasizing construction over activities, management, and utilization, and emphasizing urban areas over rural areas will be completely changed. In other words, both urban and rural public cultural spaces will be guaranteed by law. These spaces can be mainly divided into the following categories:

First, comprehensive public cultural facilities. In 2030, there will be comprehensive public cultural centers at the levels of the state, provinces, municipalities, counties, and townships. Even communities and administrative villages will have their own independent and free comprehensive public cultural stations that meet national standards. By then, China will be an aging society. There will a large number of seniors who are still in good health and have a lot of leisure time. They will become part-time or full-time staff of cultural stations. A cultural station will serve mainly as:

A library that provides books, newspapers, and periodicals, where people can read books, newspapers, and magazines. In urban areas, the books and periodicals will be mainly about general knowledge and travel information. In rural areas, they will be mainly about agriculture, rural areas, and farmers. CPC newspapers and periodicals, books and periodicals on the history of the CPC and basic local conditions, and basic public service guides will be available in both urban and rural areas. The number of books, newspapers, desks, and chairs will be provided according to the population of the village or community.

An audio-visual room, where people can listen to the radio, watch TV and movies, and surf on the Internet. By then, there will be a considerable number of videos available. Most of them will be communication and education movies introducing the history of the CPC and the basic situation of the local area, as well as the latest movies that have been out of theaters.

The audio-visual room will also be a multipurpose hall for lectures, karaoke, and mass meetings. It will be equipped with some high-definition digital TV sets and computers, projectors, and disk players, as well as high-speed Internet connection. Meetings of administrative village or community self-government organizations, video conferences with superiors, or mass mobilization and communication meetings can be held in the audio-visual room. In short, the audio-visual room will become the center of political and public cultural activities of primary-level administrative villages and communities.

A board and card game room, in which people can have non-gambling entertainment and leisure activities, such as board games, card games, and mahjong. The room is mainly for the elderly, which is an inevitable requirement for China as an aging society.

A fitness room, where people can play ball games and exercise. This is an inevitable result of the development of China's public sports services. As most people in cities will work on computers, mandatory exercise will be the main way to stay healthy. This public facility in communities and administrative villages should meet the needs of most people.

A self-study room for primary and secondary school students to study by themselves. This is the inevitable result of China's vigorous promotion of a learning society. It is also a concrete measure to build a service-oriented government. Public self-study rooms are an inevitable result of the reform of the focus of education returning to parents. The number of self-study desks will be allocated according to the number of students in the village or community.

An infants' room. It provides a "collective home" for infants and young children in the administrative village or community and public platform for the parents of these infants and young children to communicate with each other and share their parenting experience.

An instrument room, where music lovers can learn, practice, and perform music for free. It can also be used to host small concerts to enrich the public cultural life of the administrative village or community. The instrument room will be equipped with some basic instruments such as drums, gongs, and strings. Better-off administrative villages and communities can equip it with more advanced instruments, such as pianos and basses.

Second, radio and TV infrastructure. In 2030, telecommunications networks, cable TV networks, and the Internet will be integrated (the tri-networks integration). People can use TV to surf the Internet and make video calls. Moreover, there will be little delay in surfing the Internet, watching TV and movies, and listening to the radio. There are three goals in this regard:

The first goal is extending digital radio and TV to every household. With the full coverage of the network infrastructure, the goal of HD digital TV entering every household will be basically realized, and the charge will continue to drop on the current basis. Paid items will be differentiated channels under special categories for smaller groups. In some remote rural areas beyond the reach of cable TV and digital TV optical fibers, local governments will support residents with preferential policies, providing special financial subsidies for residents purchasing mobile phones or wireless network services. Residents will indirectly enjoy the public cultural services of the Extending Digital Radio and TV to Every Household Project by means of mobile Internet access.

The second goal is to make car-mounted TV a popular platform for public cultural communication. Car-mounted TV will become a popular public cultural communication platform covering urban and rural public transportation. Urban and rural buses, metros, taxis, and other public transport will be equipped with car-mounted TV. Its picture quality and sound quality will be greatly improved by then.

The third goal is human-computer interaction. With the continuous advancement of Internet technology and popularization of its applications, human-computer interaction will be available on radio, movies, and TV. People will be able to express their views on the radio and TV programs they listen to or watch through mobile phones, tablet computers, computers, TV, and other terminals anytime and anywhere, and communicate and interact with other listeners or viewers in real time. In radio and television programs with interaction parts, people can also interact with the host in real time.

Third, culture trucks. Culture trucks will become an effective supplementary carrier for basic public cultural services. In areas beyond the reach of basic public cultural services, such as poverty-stricken areas, natural villages, remote and sparsely populated areas (especially by herdsmen), and floating population areas, culture trucks will regularly bring books, movies, videos, science and technology information, small theatrical performances, small painting and calligraphy exhibitions, and entertainment equipment to meet the basic public cultural needs of local residents. In 2030, this practice will be carried out at least twice a year.

Fourth, sports infrastructure. Each administrative village and community will have an outdoor space dedicated to physical fitness, in which sports fitness equipment, such as waist twisters, parallel bars, horizontal bars, elliptical trainers, and monkey bars, as well as sports facilities, such as basketball baskets and table tennis tables will be available.

Each administrative village will have an outdoor venue dedicated to basketball, volleyball, badminton, and other mass sports activities that require a large area. In cities, communities within a 1-km radius will share a venue. Moreover, some of the sports facilities in schools will be open to the public. Local residents can use the school's sports facilities. In short, every administrative village and community must have at least one venue for the above sports activities.

Fifth, network infrastructure. In terms of the Internet, in 2030, China's Internet broadband coverage will be 100%; the computer penetration will be 80%; and the bandwidth will be greatly increased. In terms of mobile terminals such as mobile phones and tablet computers, in 2030, there will be more than one mobile phone per capita in the country. Tablet computers can be found in all urban households and over 50% in rural households. 5G mobile communication technology will be widely used, and its download speed will be almost the same as that of the broadband network.

6.4 Necessary Support Policies for the Modernization of Basic Public Cultural Services

6.4.1 Policies for Basic Public Cultural Infrastructure

Comprehensive Cultural Facilities and Sports Infrastructure
In 2030, there will be nationwide, clear, and specific unified requirements for comprehensive cultural and sports facilities. Comprehensive cultural facilities and sports infrastructure must be included in the local spatial layout planning, and given priority and emphasis from the source. At the same time, flexible policies for diversified provision by the government, NPOs, enterprises, and individuals and open bidding will be adopted. On the premise of meeting national standards, the winning NPO, enterprise, or individual will be allowed to title the facility and advertise on media and post advertisements in certain locations.

Radio, TV, and Internet Infrastructure
In 2030, the access qualification will be further relaxed. However, for national cultural security, it will be necessary for the government to set some access thresholds for radio, TV, and Internet infrastructure. The government will openly invite tenders to enterprises. After completion, the winning bidder will deliver to the government departments of information, radio, and TV for acceptance. The enterprise will operate and maintain as a franchisee. These procedures will be linked and strictly enforced. To further expand funding sources, private enterprises will be encouraged to participate, but foreign investment will still be rolled out.

The Public Digital Culture Sharing Project and its Extension to Every Household Project
Based on the Guiding Opinions on Further Strengthening the Construction of Public Digital Culture released in 2011, in 2030, there will be detailed support policies for the Public Digital Culture Sharing Project. Sound sharing libraries for public digital cultural resources will be established

across the country. There will be clear regulations and constraints on the sharing in society and among local organs, enterprises, and institutions at all levels. The Extending to Every Household Project will also have a clear schedule and quality requirements.

Culture Trucks
Regulations on culture trucks will be clearer and more specific, including provision by the government or entrusted to NPOs, funded by financial expenditures or subsidies. At the same time, there will be specific requirements for the vehicle's starting conditions, minimum hardware configuration (including the standard configuration of the vehicle and the driver's qualifications), and software configuration (including the type and quantity of movies and musical instruments that must be carried and the qualification and quantity of the staff).

6.4.2 Policies for Production

Policy for the Creation and Production of Political Cultural Goods
By 2030, public cultural goods, such as political theories on the CPC and the government, political communication, current affairs and politics, conference documents, other public events, institutional settings, and personnel information of the CPC and the government, will be produced by designated cultural institutions or enterprises. A clear division of labor will be adopted, that is, the main creators will be the CPC, the government, and their departments. The intellectual property rights will be state-owned. The main producers will be state-owned enterprises. Operations in this area will be franchising. Most of the revenue will be the government's financial expenditure for purchasing such public cultural goods.

The creation and production of performable public cultural goods to communicate the glorious history and image of the CPC and the government, CPC and state leaders, and models are special. Their main creators will be cultural enterprises with the qualifications stipulated by the state. Their broadcasting, performance, or production will depend on the specific form of the product.

The above two types of public cultural goods will be approved by the CPC committee and competent government departments before they are officially put into production, that is, the pre-production approval system.

This will not be changed significantly from the current policy requirements. However, in order to improve production efficiency, administrative approval procedures will be reduced, and the approval power will be delegated.

The public cultural goods here are public cultural products other than the above-mentioned political cultural goods. Such cultural goods can be created and produced by the government, or by NPOs, enterprises, and individuals. When the creation and production are carried out by the latter group, the government will give some policy support according to the actor, such as financial subsidies to NPOs, financial subsidies or tax concessions to enterprises, and rewards or commendations to individuals.

6.4.3 Policies for Provision

Government Procurement

It is mainly manifested in four forms:

First, franchising. The government authorizes enterprises to produce a product or use public property or enjoy the exclusive right to operate a business in a certain area according to the needs of public utilities, public safety, social welfare or the provisions of the law. In 2030, CPC newspapers and periodicals and radio and TV agencies are expected to be transformed into state-owned enterprises. Therefore, generally in this regard, the government will franchise these enterprises to produce public cultural goods and purchase the goods with fiscal expenditure.

Second, service outsourcing. Government agencies or departments outsource logistics and technology services and public services. Under an access system with clear conditions and qualifications, contractors will be publicly selected following certain procedures. Government agencies or departments will sign contracts with market entities (social organizations), whose reward will be linked to the quantity, quality, and efficiency of services and the results of standardized assessments. With the continuous streamlining of government agencies, the provision of many public cultural services has been outsourced to further improve the efficiency and effectiveness of public services.

The third is contracted commission. The government commissions enterprises to produce public cultural goods that it needs. The government shall set the standards for public cultural goods and services, and open tenders to enterprises that meet the qualifications. The winning enterprise signs a production contract with the government and produces in accordance with the contract and in compliance with law. In this process, the government must ensure some profit for the winning enterprise. Compared with the current policy, the future policy will be clearer and more specific, and public cultural goods in different areas will have different regulations, including the qualifications of bidding enterprises and the profit rate of different public cultural goods.

The fourth is direct procurement. The government directly purchases some public cultural goods produced by enterprises according to the needs of public cultural good provision. For example, the local government needs to allocate a batch of books for the city's public library and purchases books from some publishing houses.

Government Subsidy

The subsidy comes in mainly two forms:

First, subsidizing the supply side. The government provides direct financial subsidies to NPOs, enterprises, and individuals who provide public cultural goods and services for free or at a price lower than the cost, so as to further motivate them to serve public cultural undertakings.

Second, subsidizing the demand side. The government provides point-to-point subsidies to consumers who have not yet enjoyed unified public cultural services provided by governments at all levels, so as to reduce the cost of enjoying the services and ensure their basic public cultural rights. For example, rural residents and herdsmen in remote areas are subsidized for purchasing Internet services and terminal equipment.

Provision Directly by the Government

It mainly includes two types of public cultural goods and services: One is physical public cultural products, services, institutions, and facilities, such as cultural centers (stations) and museums, children's (youth's) palaces, cultural palaces, and martyrs' cemeteries. The other is the public cultural goods, services, institutions, and facilities that represent the national artistic level and the image of the CPC and China. By 2030, the two types of institutions will still be public institutions. Therefore, their provision of public cultural goods and services is essentially direct provision by the government.

Provision by NPOs, Enterprises, and Individuals

There will be special laws to encourage NPOs, enterprises, and individuals to provide public cultural goods in the following aspects:

First, the regulations on provision qualifications will be relaxed to encourage famous, influential, or featured cultural goods or collectibles and the former residences of historical and cultural celebrities to be exhibited and opened to the public free of charge.

Second, the provision scope, standards, and procedures will be regulated with other provision requirements to encourage the provision of featured folk public cultural goods, collectibles, and former residences of celebrities beyond the scope of government provision, as well as public cultural goods created and produced by enterprises. Unified supply standards will be formulated, focusing on necessary exhibition requirements, safety standards, and services that the government can provide. The supply process will be streamlined and approval procedures will be reduced.

Third, provision will be given some support and encouragement, mainly including direct financial subsidies, tax concessions, bonuses, and honors, to maximize providers' positive effects and help them reduce costs. This is conducive to the formation of a positive incentive cycle to stimulate more and more NPOs, enterprises, and individuals to provide public cultural goods.

6.4.4 Policies for Consumption

Totally Free (in Two Types)

First, free for all. This is mainly for political public cultural goods, such as CPC newspapers and periodicals; government information; healthy-themed songs, movies, and TV shows; revolution museums; memorial halls; and martyrs' cemeteries. The biggest difference between the future policy and the current policy is that CPC newspapers and periodicals and healthy-themed songs movies, and TV shows will all be free.

Second, free for key groups, including minors, the elderly, people with disabilities, active servicemen, and low-income groups. This policy will be adopted mainly by natural heritage parks, cultural heritage parks, protected historical and cultural sites, historical and cultural cities, and historical and cultural towns (villages). In 2030, more cultural entities will adopt this practice. The special offer enjoyed by key groups will be more than lower fares than ordinary people.

Basically Free

This policy applies to public cultural goods and services that basically charge only for the infrastructure costs. They mainly include public cultural goods and services with huge infrastructure costs, such as public cultural products and services based on the Internet, radio and TV, and information services. They will adopt a basically free policy that will benefit all.

Lower Than the Market Price

This policy is mainly aimed at some emerging public cultural goods and services that have just been listed as public cultural services, as well as public cultural services provided by some NPOs and individuals. In order to encourage and urge them to provide better public cultural goods and services in the long term, preference policies of prices below the market price will be adopted, and the government will subsidize them.

Progressive Low Prices

Fees will be charged differently according to the main service object and the secondary service object. For example, the art museum charges adults, senior high school students, junior high school students, and younger students a service fee from high to low and to free. The main difference between the future policy and the current policy is that the special offer classification will be further divided and the service fee will be reduced.

6.4.5 Policies for Funds

Increasing the Proportion of Expenses for Cultural Undertaking in Fiscal Expenditure

The proportion of expenses for cultural undertaking in fiscal expenditure will increase year by year. There will be a new indicator, per capita public cultural expenditure, which must increase year after year. After the 11th Five-Year Plan, the average annual growth rate of expenses for cultural undertaking should remain at a high level of over 18%,[17] and will only increase in upcoming years in the future and remain higher than the increased rate of recurrent financial revenue.

[17] http://news.xinhuanet.com/photo/18cpcnc/2012-11/07/c_123925238.htm.

Increasing Transfer Payments

Fiscal transfer payments will be increased in order to fill in the "depression" of public cultural spending as much as possible, completely changing the current inequity of emphasizing urban areas over rural areas. Transfer payments will mainly focus on:

First, rural areas, especially old revolutionary base areas, ethnic minority areas, remote mountainous areas, poverty-stricken areas, and contiguous areas of dire poverty. It is necessary to ensure the effective spending of the special funds for the Border Culture Corridor.

Second, primary-level areas, especially administrative villages, small and medium-sized cities, and towns. It is necessary to ensure that the local special funds for cultural facility repair and equipment purchase focus on the primary level.

6.4.6 Policies for Cultural Economy

The State Council has finalized some policies and measures, such as establishing a special fund for the development of communication and culture, levying the cultural undertaking fees, implementing the refund after the collection of value-added tax on communication and cultural entities, implementing five preferential policies for cinema, and encouraging donations to cultural undertakings. On this basis, the cultural financing mechanism and multi-channel investment system will be further improved.

6.4.7 Policies for Administrative Management

The "General Culture" Centralized Management Policy

Lessons will be drawn from the merger of press and publication, radio and television, culture, and other departments since the reform of the cultural system, and the integration of relevant departments of local governments at all levels will speed up. A pattern of unified administration of public culture will be established under the thinking of "general culture" to change the status quo of fragmentation and reduce administrative costs. That way, the administrative efficiency will be improved, and the coordinated construction and rapid development of the public cultural service system will be promoted to the greatest extent.

"Coordinating the Immediate Lower Rank" Policy
It mainly applies to local governments. It means that the provincial government manages the public cultural services of the province and prefecture-level governments under it; the prefecture-level government manages the public cultural services of the county (district) government; and the county (district) government manages public cultural services at the township (sub-district) level; the township (sub-district) manages public cultural services at the administrative village (community) level. This practice will effectively improve the public cultural service capabilities, the overall planning awareness, and the holistic view.

6.5 Path Arrangement for the Modernization of Basic Public Cultural Services

6.5.1 *Strategy for the Development of Basic Public Cultural Services*

Prioritizing the Development of the Public Cultural Service System
Governments at all levels should no longer just emphasize the development of public education, science and technology, and health service systems, and place equal emphasis on the development of the public cultural service system, rather than treating it as some optional "soft task." They should also increase their support for the development of the public cultural service system and really favor it in human, financial, and material resources.

Production Before Service
First, it should be expanding the production of public cultural goods according to the size of financial resources. After the production mechanism is improved, goods are enriched, and systems and mechanisms fit in, the innovation and changes can be tried on the service mechanism. Finally, the practice can be rolled out nationwide.

Public Before Political
Compared with the political sphere, the public sphere is less affected by policy changes. Therefore, cultural policies in the political sphere should be changed only after the policies in the public sphere are stable and the reform in the political sphere is basically effective.

6.5.2 Roadmap and Timetable for the Modernization of Basic Public Cultural Services

Infrastructure Projects

Infrastructure projects are the improvement of hardware, and should be designed following the principle of "in-advance as financial resources permit."

The first is to launch the upgraded version of the "Five Cultural Projects for the People" in a timely manner. Based on the now basically completed "Five Cultural Projects for the People" (the Extending Radio and TV Coverage to Villages Project, the Cultural and Information Resource Sharing Project, the Township Comprehensive Cultural Station Project, the Rural Movie Screening Project, and the Rural the Reading Room Project), the upgraded version, the "Four Cultural Projects for the People" should be launched as soon as possible. They are the Extending Digital Radio and TV Coverage to Households Project, the Cultural and Information Resource Deep Sharing Project, the Village Comprehensive Cultural Station Project, and the Rural Digital Movie Screening Projection. The goal is expected to be achieved by 2023. After 2023, support policies and their implementation rules will be formulated, supplemented, and improved.

The second is to greatly improve the speed of public digital cultural resource sharing based on the "Broadband China" strategy. The "Broadband China" strategy and implementation plan outlined the target and timetable for China to become a leader in international broadband accessibility and aims to speed up the network, lower the service fees, and reach more corners of the country. According to the plan, the sharing speed of digital public cultural resources must be higher than the above-mentioned speed. In 2030, the average broadband access capacity in urban and rural areas will be higher than 100 MB. The penetration rate of 5G will exceed 90% in cities and exceed 50% in rural areas. The "tri-networks integration" will basically be achieved.

With the above efforts, in 2030, residents across the country can enjoy most free public digital cultural resources without leaving their homes. They can enjoy modern public cultural services within their village or community.

Reform of the Cultural System

First, the institutional reform of cultural administrative departments. The Central Government is generally reappointed every five years. Correspondingly, in 2018, the Ministry of Culture and the National Tourism Administration will merge. In 2023, the new Ministry of Culture and the General Administration of Sport will merge. In 2028, the new Ministry of Culture and the State Administration of Press, Publication, Radio, Film, and Television will merge. After the three mergers, the new agency will still be called the Ministry of Culture. Institutional integration at the provincial, prefecture, and county levels will also be completed.

Second, transforming public institutions into enterprises. According to the stability period of about ten years for China's policies and the development strategy of "the social sphere before the political sphere," it is expected that by 2025, current affairs and political news programs will see the integration of production and broadcasting and the integration of interviews and editing, while other types of programs will see the separation of production and broadcasting and the separation of interviews and editing. In 2030, all publishing houses, periodicals, radio stations, TV stations, CPC newspapers, and parks as public institutions will be transformed into state-owned enterprises. Their strict division of business will be further broken, forming a large competitive market of culture across the country.

Third, non-profit cultural institutions and NPOs. In 2025, the classified management of pure non-profit cultural institutions will be further refined and standardized. Cultural institutions that are purely non-profit will adopt a public-institution system of full appropriation or partial appropriation (according to the country's needs and the number of visitors). Financial subsidies for martyrs' cemeteries, former residences of celebrities, revolutionary sites, museums, and cultural relics will be further increased. For some museums (cultural stations) with a large flow of visitors, such as youth's palaces, cultural palaces, and the Palace Museum, the financial investment will be appropriately reduced to encourage them to generate income by themselves. At the same time, the state will adopt a more relaxed policy to encourage the development of cultural NPOs. Around 2025, NPOs will enter a period of vigorous development. By 2030, there will be a considerable number NPOs for public culture.

Provision of Public Culture

Before 2025, public culture will still be dominated by government provision. After 2020, as corporate social responsibility is increasingly recognized by enterprises themselves and encouraged by the government and society, NPOs for public culture will develop by leaps and bounds. The income of the public will rise sharply, and there will be more and more public culture provided by enterprises, society, and individuals. Therefore, it is expected that in 2030, although public culture will still be dominated by government provision, other actors will become very significant. This will be reflected in developed urban areas and eastern coastal areas and sub-sectors of public cultural services.

In terms of transfer payments, the government will continue to increase transfer payments for the development of the public cultural service system in Central and Western China and rural areas. Therefore, in 2030, although the provision public culture in Central and Western China and rural areas will still be dominated by the government, the basic public cultural services will be equalized on the whole and objectively.

Consumption of Public Culture

As the government's financial resources continue to grow, in 2025, charges for public cultural goods and services will further decline. As the system for government procurement of public cultural goods continues to improve, by 2030, most public cultural goods, including CPC newspapers and periodicals and healthy-themed songs, movies, and TV programs, will be free of charge. At that time, regional public cultural goods and services will be charged in a progressive manner. The Internet access fee will be close to the current cable TV access fee.

To conclude, based on the 12th Five-Year Plan for the National Basic Public Service System, the list of basic public cultural services in 2020 and 2030 and their provision summary forecasted in this study are shown in Table 6.1.

Table 6.1 List of basic public cultural service items in 2020 and 2030

Category	*Sub-category*	*2020*					*2030*				
		Service Target	*Security Standard (Consumption Policy)*	*Expenditure Responsibility*	*Coverage*	*Provision Means*	*Service Target*	*Security Standard (Consumption Policy)*	*Expenditure Responsibility*	*Coverage*	*Provision Means*
(I) Press and Publication	1. Public cultural goods of the ruling party	All CPC members	Provided at CPC and government organs and the CPC organizations of administrative villages and communities, free for browsing and borrowing at the above-mentioned organs and their service outlets	Jointly by the Central and local governments	Fully covering target organizations	Government provision	All CPC members	Distributed for free on demand	Jointly by the Central and local governments	Covering 100% of the target group	Franchising
	2. Public information goods of governments	Urban and rural residents	Provided at CPC and government organs and the CPC organizations of administrative villages and communities, free for browsing and borrowing at the above-mentioned organs and their service outlets	Jointly by the Central and local governments	Covering 100% of residents on business	Government provision	Urban and rural residents	Distributed for free at government agencies at all levels and their service outlets and CPC village and community committees	Jointly by the Central and local governments	Covering 100% of residents on business	Franchising
	3. News information, newspaper, and periodicals on basic livelihood	Urban and rural residents	Provided at CPC and government organs and the CPC organizations of administrative villages and communities, free for browsing and borrowing at the above-mentioned organs and their service outlets	The local government, reasonably subsidized by the Central Government	Basically covering all administrative villages and communities	Government provision	Urban and rural residents	Distributed for free on demand	The local government, reasonably subsidized by the Central Government	Basically covering all households	Government procurement

4. Basic life information	Urban and rural residents	The content covering basic life information of the local province, municipality, and county; provided for free at CPC village and community committees	The local government	Basically covering all administrative villages and communities	Government provision	Urban and rural residents	The content covering basic life information of the local province, municipality, county, town, and village; provided for free at CPC village and community committees	The local government	Basically covering all administrative villages and communities	Government procurement
5. Publications in ethnic minority languages	Ethnic minority groups with their own written language	Access to common books, periodicals, and electronic audio and video products published in their own language at a reasonable price, with publication funds provided by the government	Jointly by the Central and local governments	1600 books, periodicals, and electronic audio and video products to be translated into ethnic minority languages every year	Market provision	Ethnic minority groups with their own written language	Access to common books, periodicals, and electronic audio and video products published in their own language at a low price, with publication funds provided by the government	The local government, reasonably subsidized by the Central Government	3200 books, periodicals, and electronic audio and video products to be translated into ethnic minority languages every year	The supply side subsidized by the government
6. Publications in Braille	People with visual impairment or low vision	Access to publications in Braille at a reasonable price, with publication funds provided by the government	Jointly by the Central and local governments	1.05 million copies of 2400 books and periodicals produced every year	Market provision	People with visual impairment or low vision	Access to publications in Braille at a low price, with publication funds provided by the government	The local government, reasonably subsidized by the Central Government	14 million copies of 3200 books and periodicals produced every year	The supply side subsidized by the government

(*continued*)

Table 6.1 (continued)

Category	*Sub-category*	*2020*					*2030*				
		Service Target	*Security Standard (Consumption Policy)*	*Expenditure Responsibility*	*Coverage*	*Provision Means*	*Service Target*	*Security Standard (Consumption Policy)*	*Expenditure Responsibility*	*Coverage*	*Provision Means*
(II) Radio and TV	7. Public radio and TV programs	Urban and rural residents	Over 6 central channels and 2 local channels of public radio and TV program services; program channels to be increased and digitalized	Jointly by the Central and local governments	Covering all households; digitalization covering 25% of households	Government provision	Urban and rural residents	Over 12 central channels and 4 local channels of public radio and TV program services; program channels to be increased, which have basically been digitalized	Jointly by the Central and local governments	Covering all households	Franchising
	8. Non-profit movies	Urban and rural residents	One non-profit movie on screen every half a year	The local government	Covering all residents	Service outsourcing	Urban and rural residents	One non-profit movie on screen every three months	The local government	Basically covering all residents	Service outsourcing
	9. Non-profit movie screening	Urban and rural residents	One movie screening in communities within a 1-km radius every month (administrative villages every half a month), each subsidized with RMB 400	Jointly by the Central and local governments by proportion	13.9 million movie screenings every year	Service outsourcing	Urban and rural residents	One movie screening in communities within a 0.5-km radius every month (administrative villages every 10 days), each subsidized with RMB 800	Jointly by the Central and local governments by proportion	20.8 million movie screenings every year	Service outsourcing
	10. Ethnic minority radio, movies, and TV	Major ethnic minority groups	Access to over 1 central channel and 1 local channel of radio and TV programs in their own language	Jointly by the Central and local governments	Covering areas inhabited by major ethnic minority groups such as the Tibetan, Uyghur, Mongol, Kazak, Korean, and Zhuang people	Government provision	Major ethnic minority groups	Access to over 2 central channel and 2 local channel of radio and TV programs in their own language	Jointly by the Central and local governments	Covering areas inhabited by major ethnic minority groups such as the Tibetan, Uyghur, Mongol, Kazak, Korean, and Zhuang people	Franchising

(III) Culture and Art	11. Cultural centers (stations) and museums	Urban and rural residents	Historical buildings and ruin sites charging half of the current admission and the basic items of other cultural centers (stations) and museums free of charge; open for no less than 10 months every year	Jointly by the Central and local governments by proportion	Covering all residents	Government provision	Urban and rural residents	The basic items of all cultural centers (stations) and museums free of charge; open for no less than 11 months every year	Jointly by the Central and local governments by proportion	Covering all residents	Government provision; service outsourcing; provision by enterprises, social organizations, and individuals
	12. Performances	Urban and rural residents	One cultural performance, such as music, dancing, acrobatics, or play, in communities within a 1-km radius every half a year (administrative villages every three months), each subsidized with RMB 800	Jointly by the Central and local governments by proportion	3.6 million performances every year	Service outsourcing	Urban and rural residents	One cultural performance, such as music, dancing, acrobatics, or play, communities within a 0.5-km radius every half a year (administrative villages every one and a half months), each subsidized with RMB 1600	Jointly by the Central and local governments by proportion	Over 7.2 million performances every year	Service outsourcing
	13. Conventions and exhibitions	Urban and rural residents	One static exhibition on fine arts, calligraphy and painting, handicraft, or stamps in every sub-district and county every half a year	Jointly by the Central and local governments by proportion	36,000 exhibitions every year	Government provision	Urban and rural residents	One static exhibition on fine arts, calligraphy and painting, handicraft, or stamps in every sub-district and county every quarter	Jointly by the Central and local governments by proportion	72,000 exhibitions every year	Government provision

(*continued*)

Table 6.1 (continued)

Category	*Sub-category*	*2020*					*2030*				
		Service Target	*Security Standard (Consumption Policy)*	*Expenditure Responsibility*	*Coverage*	*Provision Means*	*Service Target*	*Security Standard (Consumption Policy)*	*Expenditure Responsibility*	*Coverage*	*Provision Means*
	14. Mass cultural and art activities	Urban and rural residents	Mass cultural and art activities such as lectures, public square dancing, and calligraphy, painting, and singing contests held in every community and township from time to time	The local government	Covering all communities and townships	Government provision or organized by society	Urban and rural residents	Mass cultural and art activities such as lectures, public square dancing, and calligraphy, painting, and singing contests held in every community and administrative village from time to time, and regularly in some communities and administrative villages	The local government	Basically covering all communities and administrative villages	Government provision or organized by society
	15. Ethnic minority culture and art	Ethnic minority residents	More than one regular cultural and art activity for each ethnic minority group every year	Jointly by the Central and local governments	Covering all ethnic minority groups	Government provision or organized by society	Ethnic minority residents	More than one regular cultural and art activity for each ethnic minority group every half a year	Jointly by the Central and local governments	Covering all ethnic minority groups	Government provision or organized by society

(IV) Mass Sports	16. Sports facilities	Urban and rural residents	Public sports facilities (including schools) where conditions permit open to the public for free in the spare time of the public, with the opening hours no less than the minimum required by the provincial government, which shall extend on statutory holidays and summer and winter vacations	The local government, reasonably subsidized by the Central Government	Venues available for free accounting for 40% of the country's total	Government provision	Urban and rural residents	Public sports facilities (including schools) where conditions permit open to the public for free in the spare time of the public, with the opening hours no less than the minimum required by the provincial government, which shall extend on statutory holidays and summer and winter vacations	The local government, reasonably subsidized by the Central Government	Venues available for free accounting for 80% of the country's total	Government provision
	17. Primary-level sports meetings	Urban and rural residents	Regional mass sports meetings and special competitions such as dragon boat racing and mountain-climbing competitions hosted by the sub-district or township from time to time	The local government, reasonably subsidized by the Central Government	Covering over 50% sub-districts and townships	Government provision	Urban and rural residents	Regional mass sports meetings and special competitions such as dragon boat racing and mountain-climbing competitions hosted by the community or administrative village from time to time	The local government, reasonably subsidized by the Central Government	Covering over 50% of communities and administrative villages and 90% of sub-districts and townships	Government provision

(*continued*)

Table 6.1 (continued)

Category	*Sub-category*	*2020*					*2030*				
		Service Target	*Security Standard (Consumption Policy)*	*Expenditure Responsibility*	*Coverage*	*Provision Means*	*Service Target*	*Security Standard (Consumption Policy)*	*Expenditure Responsibility*	*Coverage*	*Provision Means*
	18. National fitness services	Urban and rural residents	Access to free fitness instructions, fitness activities, and fitness knowledge; fitness equipment provided for free in public venues such as parks and green spaces	The local government, reasonably subsidized by the Central Government	Frequent exercisers accounting for over 40% of the total population	Government provision	Urban and rural residents	Access to free fitness instructions, fitness activities, and fitness knowledge; sports NPOs and instructors available; fitness equipment provided for free in public venues such as parks and green spaces	The local government, reasonably subsidized by the Central Government	Frequent exercisers accounting for over 62% of the total population; the coverage of NPOs and instructors extending to communities and administrative villages in developed areas and to sub-districts and townships in other areas	Government provision
(V) Internet and Information(I) (I)	19. Internet access services	Urban and rural residents	Internet service rates to drop by a half based on the current price		Covering 100% of the target group	Market provision	Urban and rural residents	Internet service rates to further drop to the current rate of cable TV	The local government, reasonably subsidized by the Central Government	Covering 100% of the target group	The supply side subsidized by the government

	20. Internet access in public places	Urban and rural residents	Free Wi-Fi services at government organs, public institution offices, and public transport	Jointly by the Central and local governments	Covering 100% of residents on business	Government provision	Urban and rural residents	Free Wi-Fi services at government organs, offices of state-own enterprises and public institutions, parks, and tourist attractions	Jointly by the Central and local governments	Covering 100% of residents on business	Government provision; provision by enterprises for free
	21. Digital form of public cultural goods	Urban and rural residents	More public cultural goods digitalized and provided for free on the Internet	Jointly by the Central and local governments	Digitalization of over 50%	Government provision	Urban and rural residents	Most public cultural goods digitalized and provided for free on the Internet	Jointly by the Central and local governments	Digitalization of over 90%	The supply side subsidized by the government
(VI) Tourism and Leisure	22. Parks, and tourist attractions	Urban and rural residents	Minors, the elderly, active servicemen, people with disabilities, and low-income groups admitted into parks and tourist attractions for free (the general admission ticket) some parks and tourist attractions open to the public for free	The Central and local governments taking their own share	Covering 100% of the target group; free parks and tourist attractions accounting for 20%	Government provision	Urban and rural residents	Minors, the elderly, active servicemen, people with disabilities, and low-income groups admitted into parks and tourist attractions for free (all tickets), some parks and tourist attractions open to the public for free	The Central and local governments taking their own share	Covering 100% of the target group; free parks and tourist attractions accounting for 50%	Government provision
	23. Board games and card games	Urban and rural residents	Venues and facilities dedicated to board games and card games in communities and administrative villages	The local government		Government provision or organized by society	Urban and rural residents	Indoors venues and facilities dedicated to board games and card games in communities and administrative villages	The local government	Basically covering all communities and administrative villages	Government provision or organized by society

(continued)

Table 6.1 (continued)

Category	*Sub-category*	*2020*					*2030*				
		Service Target	*Security Standard (Consumption Policy)*	*Expenditure Responsibility*	*Coverage*	*Provision Means*	*Service Target*	*Security Standard (Consumption Policy)*	*Expenditure Responsibility*	*Coverage*	*Provision Means*
(VII) Infrastructure	24. Comprehensive public cultural facilities	Urban and rural residents	Comprehensive public cultural centers (stations) in sub-districts and townships	The local government, reasonably subsidized by the Central Government	Basically covering all sub-districts and townships	Government provision	Urban and rural residents	Comprehensive public cultural centers (stations) in communities and administrative villages	The local government, reasonably subsidized by the Central Government	Basically covering all communities and administrative villages	Government provision or provision by enterprises for free
	25. Radio and TV infrastructure	Urban and rural residents	Digital TV network available in communities and administrative villages; car-mounted TV which can broadcast public cultural information in real time popularized in large and medium-sized cities	Jointly by the Central and local governments	Basically covering all communities and administrative villages; car-mounted TV covering large and medium-sized cities	Franchising	Urban and rural residents	Tri-network integration achieved; digital TV network extended to every household; car-mounted TV which can broadcast public cultural information in real time widely applied on metros, taxis, and urban-rural shuttle buses	Jointly by the Central and local governments	Basically covering 100% of the target group	Franchising

26. Culture trucks	Urban and rural residents	Providing free mobile cultural services such as movie and TV screening, reading, entertainment utensils, small artistic performances, and calligraphy and painting exhibitions, at least once a year	The local government, reasonably subsidized by the Central Government	Basically covering natural villages, remote areas, and migrant workers' settlements that are beyond the reach of the above-mentioned services	Government provision	Urban and rural residents	Providing free mobile cultural services such as movie and TV screening, reading, entertainment utensils, small artistic performances, and calligraphy and painting exhibitions, at least twice a year	The local government, reasonably subsidized by the Central Government	Basically covering natural villages, remote areas, and migrant workers' settlements that are beyond the reach of the above-mentioned services	Government provision
27. Sports infrastructure	Urban and rural residents	Sports and fitness venues and facilities in all communities and administrative villages	The local government, reasonably subsidized by the Central Government	Basically covering all communities and administrative villages	Government provision	Urban and rural residents	Sports and fitness venues and facilities fit for popular ball games in all administrative villages; such venues and facilities to be found in a 1-km radius of each community	The local government, reasonably subsidized by the Central Government	Basically covering all communities and administrative villages	Government provision; service outsourcing; provision by enterprises, social organizations, and individuals

(*continued*)

Table 6.1 (continued)

Category	*Sub-category*	*2020*					*2030*				
		Service Target	*Security Standard (Consumption Policy)*	*Expenditure Responsibility*	*Coverage*	*Provision Means*	*Service Target*	*Security Standard (Consumption Policy)*	*Expenditure Responsibility*	*Coverage*	*Provision Means*
	23. Network structure	Urban and rural residents	Internet access and Internet terminal devices such as PC or tablet computers available in all communities and administrative villages	The local government, reasonably subsidized by the Central Government	Internet basically covering all communities and administrative villages; PC covering 100% households in cities and 20% in rural areas; tablet computers covering 50% households in cities and 10% in rural areas	Market provision	Urban and rural residents	Internet access and Internet terminal devices such as PC or tablet computers available in all communities and administrative villages; bandwidth greatly expanded	Jointly by the Central and local governments	Internet basically covering all communities and administrative villages; PC covering 100% households in cities and 80% in rural areas; tablet computers covering 100% households in cities and 50% in rural areas	The demand side subsidized by the government

CHAPTER 7

Modernization of Basic Social Services

7.1 Connotation of the Modernization of Basic Social Services

7.1.1 Basics of Social Services

As defined by the Ministry of Civil Affairs of the People's Republic of China, social services refer to the policy system and institutional arrangements led and implemented by the government to provide chore help and care services and support for citizens and their families, especially socially disadvantaged groups and special groups (such as the elderly, people with disabilities, children, the unemployed, and the poor), to secure their survival and development rights and meet their need for a dignified life in the process of modernization. The social services led by the Ministry of Civil Affairs mainly include social assistance, social welfare, elderly care services, and preferential treatment to entitled groups.

Social services should be said to be an institutionalized means and approach adopted by the state to promote the rational allocation of social resources in response to social problems and possible contradictions arising from huge social changes and social transformation. They are the outcome of human society modernization and care and respect for the basic human needs. Their level directly reflects the degree of social civilization.

In China, social services are integral to the social security system and fundamental to the long-term stability of the country. The Decision of the

K. Dong, N. Wei, *Transforming China's Public Services*,
https://doi.org/10.1007/978-981-99-3943-5_7

CPC Central Committee on Several Major Issues Concerning Building a Socialist Harmonious Society expounds on social services as "focusing on enhancing the functions of social services and improving the capabilities of social management and administration of law, and vigorously strengthen primary-level governance." Specifically, it was proposed to "improve the public service system and ensure people's basic quality of life. Social services should adapt to the aging of the population, urbanization, and diverse employment. A social security system should be established for urban and rural residents where social insurance, social assistance, social welfare, and charity are connected with each other." The outline of the state's 12th Five-Year Plan called on "adhering to the guidelines of 'broad coverage, basic insurance, multiple levels, and sustainability,' accelerating the construction of the social security system for urban and rural residents, and steadily improving the level of security." Its proposition on social services was to "strengthen the construction of the social assistance system" and "actively develop social welfare and charity undertakings."

In recent years, as social development and people's livelihood security attract more attention, the state has raised further requirements on the coverage, security level, operation management, and provision efficiency of social services. The 18th National Congress of the CPC called on "coordinated development of the social security system in urban and rural areas" and "improving the social assistance system and the social welfare system, supporting the development of charity, and ensuring preferential treatment to entitled groups." The 13th Five-Year Plan further proposed to reform and improve the social security system, "adhere to covering the entire population in both urban and rural areas with support that hits the right level, rights and responsibilities clearly defined, and efficient operation, steadily improve the overall level of social security, and establish more equitable and sustainable social security systems."

This indicates two significant attributes of social services in China's policy system. First, social services are part of the social security system. They now focus on providing basic living security for groups in difficulties and special groups. Second, many social services are given the function of strengthening primary-level governance.

7.1.2 *Definition of Basic Social Services*

According to the 12th Five-Year Plan for the National Basic Public Service System, basic public services are "provided based on social consensus, led by the government, adapted to the level and stage of economic and social development, and aimed at meeting the basic needs of all citizens to survive and develop. Receiving basic public services is a right of citizens, and the provision of basic public services is the responsibility of the government." The "basic" emphasizes civil rights and government responsibilities.

In this sense, basic social services can be understood as provided under the leadership of the government to secure citizens' rights of a dignified life and equal participation in social development. The 12th Five-Year Plan for the National Basic Public Service System, which defines the main content of current basic social services, points out that "the state establishes a basic social service system to provide physical assistance for the basic living of urban and rural residents, especially disadvantaged groups, and ensure a dignified life and equal participation in social development for special groups such as the elderly, people with disabilities, and orphans."

7.1.3 Connotation of the Modernization of Basic Social Services

Modernization is the common dream of the Chinese people. It conveys the people's anticipation for social life and cultural changes and envisages highly developed production, highly enriched well-being, and highly advanced civilization. The modernization of public services is relative. It is reflected in that public services "better respond to citizens' needs, have a higher level of security, are provided in a more equitable, convenient, and efficient manner."

Based on this notion, modernized basic social services should have three connotations. First, basic social services seamlessly meet the needs of all citizens for a dignified life and equal participation in social development. Public participation is emphasized in the decision-making and social service provision. Second, the provision of basic social services is compatible with the level of economic and social development, and the content and level meet the growing common needs of the people. Third, residents have easy access to equalized basic social services, and the efficiency of government resource investment continues to improve.

7.2 Current Status of and Problems in Basic Social Services

7.2.1 Current Frame of Basic Social Services

At present, the structure, system construction, and provision of basic social services in China are led by civil affairs authorities. The "basic social services" in the current national policy mainly include social assistance, social welfare, preferential treatment to entitled groups. Civil affairs authorities have corresponding institutions for these functions. In the context of population aging, civil affairs authorities are paying more and more attention to basic elderly care services. When setting the basic standards for basic social services to be achieved during the 12th Five-Year Plan period, the 12th Five-Year Plan for the National Basic Public Service System lists basic elderly care services as one of the four major areas.

Specifically, social assistance is mainly for people with low income, suffering from natural disasters, or having lost the labor ability, which provides them with physical or mental assistance to meet their basic living needs and ensure their minimum living standards. For a long time, social assistance has mainly included minimum living security, rural Five Guarantees (the state guarantee on proper food, clothing, medical care, housing, and funeral expenses) provision, medical assistance, relief for vagrants and beggars, and temporary assistance. Minimum living security is divided into urban and rural according to household registration and economic development level. Relief for vagrants and beggars is divided into relief for vagrants and beggars without assured living sources in cities and relief for minor vagrants. Temporary assistance is divided into natural disaster assistance, temporary assistance for families in difficulties due to illness, and staged life assistance for low-income families.

In February 2014, the State Council released the Interim Measures for Social Assistance, which collates and includes the content of social assistance. It clearly features minimum living security, support for people living in dire poverty, assistance for disaster victims, medical assistance, assistance for schooling, housing assistance, employment assistance, and temporary assistance. The introduction of the Interim Measures is an important measure for the national system to promote the institutionalization of social assistance and expands the extension of social assistance.

Social welfare is mainly for disadvantaged and special groups, including children, the elderly, and people with disabilities. Specifically, children's welfare was mainly about orphan raising in the early stage. It is now incorporating the raising of de facto unattended children, that is, children with AIDS, severe illness, severe disabilities, and rare diseases, and children of criminals in custody or prison. Seniors' welfare includes the allowance (subsidy) system for the elderly, which has been basically established, and the assistive device provision system for seniors with disabilities and semi-disabilities, which is being established. The welfare of people with disabilities includes the provision of special education, employment, culture, sports, and other services for them.

Preferential treatment to entitled groups is a basic social service led by civil affairs authorities mainly for active servicemen of the People's Liberation Army of China, revolutionary veterans with disabilities, demobilized servicemen, and families of revolutionary martyrs, servicemen who have died in service or illness, and active servicemen. The treatment mainly includes (1) supporting the army and giving preferential treatment to the families of servicemen and martyrs; (2) providing pensions, subsidies, and preferential treatment to families of martyrs, servicemen with injuries or disabilities, veterans of the Red Army living in rural areas, demobilized serviceman living in rural areas, demobilized serviceman returning home with illness, and family members of active servicemen; (3) providing casualty pensions to state organ workers, policemen, militia, and migrant workers engaging in battles; (4) approving and commending martyrs; (5) establishing rehabilitation hospitals for servicemen, chronic disease nursing homes for demobilized servicemen, mental hospitals for demobilized servicemen, and glory homes; and (6) managing and preserving memorial buildings for martyrs and compiling records of martyrs' deeds.

In recent years, civil affairs authorities carry out some work similar to basic social services in the marriage registration management and the funeral reform, which it is responsible for, such as the trial of marriage counseling and providing funeral subsidies to those who do not retain ashes and the deceased with low incomes in areas where cremation is implemented. These practices are being incorporated into the basic social service system by civil affairs authorities.

In fact, the maternity services and health services currently progressed by the family planning administration have the characteristics of basic social services and are related to national policies. The Disabled Persons' Work Committee of the State Council and China Disabled Persons' Federation specially established by the state also play an important role in basic public services for people with disabilities.

7.2.2 *Analysis of Current Operation of Basic Social Services*

As far as social assistance is concerned, minimum living security, natural disaster assistance, medical assistance, living relief for vagrants and beggars, and relief for and protection of minor vagrants have basically covered all their target groups, and temporary assistance for families in difficulties due to illness and staged life assistance for low-income families are still under progress. According to the announcement of the National Audit Office of the People's Republic of China, by the end of 2011, there were 22.5627 million recipients of minimum living security for urban residents, 52.9828 million recipients of minimum living security for rural residents, and 5.7862 million recipients of rural Five Guarantees. Urban and rural medical assistance funds subsidized the medical insurance coverage of 66.4935 million people and provided direct assistance after medical treatment for 23.6727 million patients.

In terms of social welfare, orphan raising has achieved comprehensive coverage. Most healthy orphans are raised well by being adopted and other means. Most orphans with disabilities are cared for at orphanages. The raising of de facto unattended children is in its infancy and has not achieved full coverage of the target group. In terms of seniors' welfare, the basic elderly care service subsidy for seniors in difficulties and the allowance (subsidy) for seniors over 80 have been basically established. The provision of assistive devices for seniors with disabilities, social elderly care services, and facility construction support has been included in the planning of civil affairs authorities. In the welfare of people with disabilities, social security and basic services have fully covered the target group. In the basic services for people with disabilities, special education at the compulsory education stage, rescue rehabilitation for children with disabilities, and employment, culture, and sports and fitness services have not fully covered the target group. Living allowances for impoverished people with disabilities, subsidies for nursing care and assistive devices for people with severe disabilities, and accessibility construction are in the preliminary research and exploration stage. By the end of 2011, there were 42,500 social welfare institutions nationwide with 3,210,500 beds, which took care of 2,379,200 seniors, people with disabilities, and children. A total of 628,000 people with disabilities were employed by social welfare enterprises.

In terms of marriage services, in addition to basic marriage registration and related services, civil affairs authorities are seeking to provide marriage counseling as a social service, in the hope to improve the quality of marriages for Chinese citizens and maintain social stability. In funeral services, funeral subsidies are provided. Basic funeral service fees are reduced or exempted and cremation funerals are subsidized to encourage the public to choose cremation and green burial.

At present, family planning services mainly include family planning health services and family planning rewards and assistance. Family planning health services include technical guidance and consultation, clinical medical services, and reproductive technology services. Family planning rewards and assistance have been sound. They include rewards for parents of only one child, rewards and assistance for families observing family planning, and special assistance for families observing family planning (families who have lost their only child).

In general, China's basic social services see increasing items, richer content, and improved quality. However, due to historical and institutional factors, basic social services confront some problems in the course of development, mainly including:

Unclear Policy Framework for Basic Social Services

Many policies and papers have defined the content of social assistance and social welfare. However, the definition is unsystematic and illogic. It is more like in the continuation of the traditional division of responsibilities within government functional departments. For example, in the social welfare work currently defined by civil affairs authorities, "orphan raising" is logically classified as social assistance. Another example: In the "basic elderly care services" listed separately in the 12th Five-Year Plan for basic public services, the "subsidy for elderly care services for seniors with disabilities or semi-disabilities and unable to take care of themselves from households with financial difficulties" is regarded as a type of social assistance, while the "subsidy for seniors above 80" is included in social welfare. More typically, the rural Five Guarantees provision is classified as "social assistance" in the 12th Five-Year Plan for the Development of Civil Affairs but as "social welfare" in the 12th Five-Year Plan for the National Basic Public Service System. In addition, available social services, such as preferential treatment to entitled groups, funeral subsidies, and family planning subsidies, are fundamentally different from social assistance and social welfare from the perspectives of guiding policy and policy orientation. Figure 7.1 shows the system of current basic social service items in China.

On the whole, as the state continues to increase investment in basic social services, basic social service items grow in a considerable number in recent years. However, these items have not been effectively collated and integrated according to their objectives and nature, which greatly affects the realization of their objectives. From the perspective of primary-level

Basic social services	Civil affairs authorities	Social assistance	Minimum living security	Urban minimum living security	
				Rural minimum living security	
			Rural Five Guarantees provision		
			Medical assistance		
			Relief for vagrants and beggars	Relief for vagrants and beggars without assured living sources in cities	
				Relief for minor vagrants	
			Temporary assistance	Natural disaster assistance	
				Temporary assistance for families in difficulties due to illness	
				Staged life assistance for low-income families	
		Social welfare	Children's welfare	Orphan raising	
				Raising of de facto unattended children	
			Seniors' welfare	Basic elderly care service subsidy for seniors in difficulties	
				Allowance (subsidy) for seniors over 80	
				Provision of assistive devices for seniors with disabilities	
				Social elderly care services and facility construction support	
		China Disabled Persons' Federation	Welfare of people with disabilities	Social security for people with disabilities	Subsidy for social insurance premium of people with severe disabilities and impoverished people with disabilities
					Basic medical rehabilitation of insured people with disabilities
					Living allowance for impoverished people with disabilities
					Subsidies for nursing care and assistive devices for people with severe disabilities
				Basic services for people with disabilities	Special education in the compulsory education state
					Education aid for impoverished people with disabilities
					Emergency rehabilitation for children with disabilities
					Employment services for people with disabilities
					Cultural services for people with disabilities
					Sports and fitness services for people with disabilities
				Accessibility construction	

Fig. 7.1 The system of current basic social service items in China

			pensions, subsidies, and preferential treatment		
		Preferential treatment to entitled groups	Centralized support for key entitled groups		
			Settlement of demobilized servicemen		
		Marriage services	Marriage counseling		
		Funeral services	Funeral subsidies		
	Family planning authorities	Family planning health services	Technical guidance and consultation		
			Clinical medical services		
			Reproductive technology services		
		Family planning rewards and assistance	Rewards for parents of only one child		
			Rewards and assistance for families observing family planning		
			Special assistance for families observing family planning		

Fig. 7.1 (continued)

practice, some policies cannot be implemented or leave room for corruption. It is precisely because the content of social service items in these fields is complicated and insufficiently connected, making it difficult to implement or supervise.

Institutional Inequity Caused by the Urban-Rural Gap and Regional Differences

Due to the urban-rural dual structure, many areas of basic social services feature significant urban-rural differences and institutional inequity. For example, there are significant differences in minimum living security between urban and rural residents. As far as the national average is concerned, in 2010, the average standard of minimum living security for urban areas was CN¥251.2 per person per month, while the standard for rural areas was CN¥117 in the same period. Although the difference exists for some reason, the minimum living security system, as a system to ensure the survival of extreme-low-income people, should be gradually adjusted so that its security level is be only linked to the level of local economic and social development, instead of being distinguished between urban and rural areas. As the Interim Measures for Social Assistance and the Opinions of the State Council on Further Promotion of Reform of the Household Registration System have been issued, the state is attempting to eliminate urban-rural differences at the institutional level. However, in practice, there are still significant differences in the sources of income, income structure, and income verification methods for urban and rural residents. There is still a lack of specific practical paths to eliminate the inequity of urban and rural minimum living security standards.

At the same time, departments and systems differ greatly in the professionalism for the investment in basic social service items. Some departments and systems of public service items fail to take regional differences into account, resulting in unsatisfactory policy implementation in some regions. For example, in practice, the minimum living security system often sets the target before deciding the beneficiary object. Such a policy is often twisted into an "average security" policy in generally impoverished areas.

Supply Inefficiency Due to Departmental Segmentation

At present, the major body engaged in basic social services is civil affairs authorities. The family planning department, management and service agencies for people with disabilities, and relevant people's organizations that undertake government functions also lead some basic social services. Under the current governance system in China, due to the widespread existence of "project-based" and "campaign-style" governance, departments usually fight for resources in the hope to expand their powers, leading to salient fragmentation. To a large extent, this has resulted in the overlapping of some functions and the absence of others among departments, who fail to cooperate to build a social service system.

At present, the content of basic social services is very complex. Nonetheless, careful review shows that a large part of it can be integrated through the convergence of policies between departments. For example, the flexible "medical assistance" in social assistance can be replaced by improving the medical insurance system and the subsidy system for low-income groups. Many assistance and welfare measures for people with disabilities can be integrated into the existing social assistance and social welfare system. In the provision of public service items, fragmentation easily leads to insufficient investment of resources in some areas and duplication of resources in others, resulting in low efficiency of public service provision.

A Low Overall Security Level with a Distinct "Targeted" Approach

The low overall security level is first reflected in single social assistance items. At present, services provided by social assistance are mainly the minimum living assistance, while education, medical, and housing assistances see no substantial progress. This single system is prone to some negative impacts. For example, the current minimum living security is mainly "filling gaps" according to the standard, which does not help the recipients to strengthen their motivation to improve themselves and actively seek employment. Second, current basic public services seriously lag behind the level of economic development. Finally, basic elderly care subsidies, funeral subsidies, and other policies still have a low coverage, without even covering all target groups. In social welfare, orphanages are still insufficient in the adoption capacity. Hospitalization and treatment services for people with mental illness are inadequate. Basic social services for children, the elderly, and people with disabilities have fully covered the target groups.

At the same time, most basic social services are still assistance, and welfare items make up just a small proportion. In other words, basic social services are mainly "filling gaps." With economic growth, basic social services should be more inclusive and see an increase of welfare items.

7.3 Outlook for the Modernization of China's Basic Social Services in 2030

In 2030, China's basic social service system will be more improved, and the modernization of basic social services will be completed on the whole. Compared with today, basic social services will better respond to the needs of citizens, with a higher security level, a sounder government-led and public-engaging mechanism, and more equitable, convenient, and efficient provision.

7.3.1 China's Basic Social Service Structure in 2030

In 2030, the basic structure of China's basic social services (see Fig. 7.2) will be clearer, mainly including three areas, namely, assisting basic social services, strategic basic social services, and welfare basic social services.

Assisting Basic Social Services

Assisting basic social services are mainly for disadvantaged groups, with social stability as the core objective, and humanitarianism as the institutional principle for this objective. In terms of the main institution, assisting basic social services are mainly reflected in a sound social assistance system.

	Basic social services		
Disadvantaged groups	Assisting	Social stability	Humanitarian
Specific groups	Strategic	Social development	Fair
All residents	Welfare	Cultural progress	Equitable
Target Group	Main Institution	Core Objective	Institutional Orientation

Fig. 7.2 Basic structure of China's basic social services

The object of assisting basic social services mainly includes unattended children, low-income groups with no working ability, low-income households, people with disabilities, and disaster victims. A system should be established for raising de facto unattended children, and special funds should be set up to ensure their education and living. For low-income groups with no working ability, a minimum living security system should be established, which covers their clothing, food, housing, medical care, and funeral. Urban and rural subsistence allowance recipients, rural Five Guarantees recipients, and key entitled groups will be the first to be included in this area. For assistance for low-income households, systems for children's education, social insurance subsidies, housing assistance, and judicial assistance should be established. A subsidy system for assistive device provision should be established for people with disabilities (the mobility impaired, excluding the elderly and other people with disabilities). Temporary minimum living allowances and disaster subsidies should be provided for disaster victims.

Strategic Basic Social Services

Strategic basic social services take social development as the core objective, and fairness is the institutional principle for this objective. Strategic basic social services are mainly for specific groups, with the preferential treatment and reward system as the core. Specific groups for preferential treatment and rewards include current "entitled groups," households who support the national family planning policy, households who support the national funeral policy, and those who act bravely for a just cause.

Welfare Basic Social Services

Welfare basic social services focus on developing the social welfare system. They are for all citizens, with equality as the core institutional principle and cultural progress the main objective. For a long time, China's welfare system has been mainly to fill gaps for the elderly, children, people with disabilities, and other special groups. By 2015, these groups have been basically covered by social welfare. In the future, efforts should be made to ensure that social welfare reaches these special groups in a fair manner and reaches more groups. From 2020 to 2030, the development of the social welfare system should mainly focus on promoting inclusive social welfare and provide more welfare benefits for all citizens.

Table 7.1 shows details of China's basic social service system and its institutions in 2030.

7.3.2 Support System for the Modernization of Basic Social Services

Ideological Support

The idea of humanitarian, reasonable, and equitable basic social services must be truly established. Basic social services will eventually be available to all citizens, with inclusiveness highlighted. From this perspective, it is necessary to break the traditional class-based and group-based discrimination as well as the urban-rural dual structure. Urban residents, rural residents, and workers enjoy basic social services and the fruits of development. The modernization of basic social services means more than the basic living guarantee for all citizens. It will allow people to live with dignity and provide them with opportunities and ways to strive for a better living standard and quality of life.

Institutional Support

Undoubtedly, the modernization of basic social services requires a variety of institutional supports including laws and policies. The key to institutional construction is to truly clarify the core objectives, target groups, and main content of basic social services, and more importantly, the interrelationships of basic social services. As a result, the development of the basic social service system should be linked more closely to the living needs of residents and the government's provision capacity. The boundaries between departments should be weakened. A science-based and efficient system with reasonable division of responsibilities and cooperation between government departments should be formed. At the same time, the financial investment mechanism should be reformed, with the expenditure responsibilities on basic social services clearly defined. The public

Table 7.1 China's Basic Social Service System in 2030

Area	*Main system*	*Core objective*	*Institutional principle*	*Main content*	*Recipient*		*Specific system*
Basic social services	Assisting basic social services	Social stability	Humanitarian	Social assistance	Disadvantaged groups	Unattended children	Raising unattended children
						low-income groups with no working ability	Minimum living security
						low-income households	Education aid for children from low-income households Subsidy for social insurance Housing assistance Judicial assistance
						people with disabilities	Subsidy for assistive device provision
						Disaster victims	Temporary minimum living security Disaster allowance
	Strategic basic social services	Social development	Fair	Social awards	Specific groups	Entitled groups	Preferential treatment Settlement of demobilized servicemen
						Households who support the national family planning policy	Subsidy for households observing family planning
						Households who support the national funeral policy	Funeral subsidies
						People who act bravely for a just cause	Subsidy for people who act bravely for a just cause
	Welfare basic social services	Cultural progress	Equality	Social welfare	All residents	People with disabilities	Accessibility construction Special education Support for services for people with disabilities
						The elderly	Subsidy for seniors over 80 Elderly care service support
						Minors	Social protection system
						Youth	Marriage counseling
						General residents	Funeral services Festival subsidies ...

finance system should also be reformed, and the transfer payment system should be improved to ensure that governments at all levels have financial rights corresponding to their powers. A sound regulatory mechanism for social services should be established to increase the transparency of social service income, expenditure, and provision.

Information Support

The modernization of the basic social service system must rely on information technology. First, it is necessary to integrate the basic data of social services, develop homologous basic data with clear authority delegated to each department and common use and maintenance of all departments, and strengthen mutual supervision between government departments. Second, it is necessary to establish a residents' social credit evaluation and management system, and make it an important basis for residents' access to basic public services, and promote cultural progress.

Spatial Support

For residents, access to basic public services, including basic social services, must be convenient. Among them, the acquisition of many basic social services requires spatial convenience, which determines that it is necessary to strengthen and improve community construction to form a spatial vehicle of basic public services. First, it is necessary to pay more attention to the social needs of people in the planning of community space, and the design of public space should meet the needs of people for rest, social interaction, and the development of neighborhood relations. Second, in the construction of community public facilities, the configuration and adjustment of facilities should match the social characteristics of community residents. Finally, under the premise of establishing an effective supervision system, more areas of basic social services should be open to social capital. Community social organizations with Chinese characteristics should be developed and incubated. The social service provision model should be innovated. That way, a highly autonomous and multi-party-engaging community governance pattern will be formed.

CHAPTER 8

Chinese Government Purchase of Public Services

8.1 Innovation of Public Service Provision in Changes

Public services are a special kind of public goods. They are non-competitive and non-exclusive goods of public value. Therefore, they cannot be provided solely relying on a market mechanism but with in-depth intervention by the government, which, however, is not equal to full intervention or direct provision by the government. In the 1960s, Elinor and Vincent Ostrom from the United States proposed to distinguish between the "provision" and "production" of public services. "Provision" refers that the government decides the appropriate type of services and the level of provision through taxation and spending decisions, and arranges and monitors the production. "Production" refers to the conversion of inputs into outputs.[1] When distinguishing between "management" and "governance," Donald Kettl pointed out that the role of government has changed, which is engaged less in producing public goods and services but more in supervising practical agents of public services.[2] Since the 1980s,

[1] Wang Puqu, Lester M. Salamon, et al., *Outsourcing Government-Financed Social Services to Civil Society Organizations: Lessons from China and abroad* (Beijing: Peking University Press, 2010), 34.

[2] H. Frederickson, *The Spirit of Public Administration*, trans. Zhang Chengfu, Liu Xia, Zhang Zhang, and Meng Qingcun (Beijing, China Renmin University Press, 2013), 54.

K. Dong, N. Wei, *Transforming China's Public Services*,
https://doi.org/10.1007/978-981-99-3943-5_8

Western countries have launched the New Public Management (NPM)-oriented reform of government administration. It features the introduction of a competition mechanism for traditional government-led public services, where the government's functions as a "provider" and "producer" of public services are stripped off, and non-governmental players are introduced to efficiently allocate resources and improve the quality of public services. The idea of relatively separating the provision and production of public goods and services means that there should be a reasonable distinction between "providers" and "producers" in public service provision. It also means that there can be diverse public services providers. The government can directly produce public services and provide them, or entrust non-governmental players to provide public services through entrustment, contracting, and procurement. At present, government purchase has become the main form of government provision of public services in developed countries and regions.

Government purchase can be defined as that some of the public services directly provided by the government are entrusted to qualified non-governmental players on the market according certain methods and procedures. The government pays the private players according to the quantity and quality of the services they provided. Simply put, it is a mechanism in which the government provides funds, social organizations contract services, and the two parties establish a contractual relationship to achieve specific public service goals. It is essentially the contractual provision of public services.[3] Government purchase, more commonly known as "contract outsourcing" in Western countries, started in the 1980s. At that time, the government agencies of some Western countries were huge, bloated, and inefficient. The pressure on fiscal expenditure continued to increase, and social welfare policies were unsustainable. The public lost confidence in the government's competence, and the government faced a serious financial, administration, and trust crisis. To overcome the crisis, the countries initiated the NPM-oriented government administrative reform.[4] Different from the passive reform in Western countries forced by financial pressures, the Chinese government purchase of public services is more an active response in the transformation of government functions.

[3] Su Ming, Jia Xijin, Sun Jie, and Han Junkui, "Chinese Government Purchase of Public Services," *Public Finance Research*, no. 1 (2010): 9–17.

[4] Yu Guo'an, "Analysis and Policy Suggestions on Government Purchase of Public Services," *Review of Economic Research*, no. 46 (2011): 36–41.

The initial direct driving force behind the marketization of public services in the United States mainly comes from the pressure of the severe fiscal deficit under economic "stagflation" and the increasing demands of citizens for public services. Differently, that in China mainly comes from changes in the system—the transition from a planned economy to a market economy.[5] In general, the Chinese government purchase of public services is mainly motivated by four needs.

8.1.1 The Need to Transform Government Functions and Build a Service-Oriented Government

Public service innovation is driven by the transformation of government functions. There would be hardly any demand for reform and innovation in public services without an awareness of the urgency and necessity of transforming government functions. Since it was first proposed in China's second institutional reform in 1988, the transformation of government functions has been the core and key of China's administrative structural reform. The core of transforming government functions lies in clarifying the boundaries between government and market, and between government and society. The first step is to define government functions correctly and reasonably, which requires the government to transform from an omnipotent government under the planned economy to a limited-liability government. By purchase of public services, the government has changed from a direct provider of public services to a maker of public service policies, a purchaser and supervisor, realizing the return of social power and the transformation of the government's role.[6] Therefore, in the critical period of comprehensively deepening the reform, it becomes necessary to vigorously promote government purchase of services and establish and improve the system for government purchase of service for appropriately handling the relationship between the government, the market, and society; catalyzing the transformation of government functions; preventing omnipotence; promoting the separation of government functions from institution functions, and the government from society; and building a service-oriented government.

[5] Yang Liqiong, "The Comparison of Public Service Marketization Between China and America and Its Implications," *Chinese Public Administration*, no. 7 (2011): 66–69.

[6] Liu Qingyuan and Wen Yingna, "Institutional Demands in the Government Purchase of Social Work Services," *Journal of Social Work*, no. 11 (2007):11.

8.1.2 *The Need to Address Supply-Demand Contradiction and Satisfy Diverse Demands for Public Services*

As the socialist market economy has been established and developed rapidly, the people's demand for public services is growing, with enriched content and higher standards. The demand of public services is diverse, multi-leveled, and complex, while the provision of public services is still dominated by the government. On one hand, it is the increasingly diverse needs of public services in the socialist market economy; on the other hand, it is the inability of the government to meet such needs. The inherent contradiction between the two drives the government to seek to improve its provision of public services and progress the reform of the public service provision mechanism, so that there will be diverse providers and provision means to meet the people's increasing demand for public services.

8.1.3 *The Need to Encourage Non-governmental Players and Improve the Quality and Efficiency of Public Service Provision*

Due to its defects in costs, efficiency, and professionalism, the government providing services as the only player will result in high costs, low efficiency, and average quality. A competition mechanism established in public service provision through purchase can guide non-governmental players to participate in public services. Entrusting public services to non-profit organizations, enterprises, or other non-governmental players can fully exert their flexible mechanisms, relevance to the people, and advantages in human resources, professional technology, and service costs. It is conducive to effectively mobilizing non-governmental players, integrating and utilizing social resources, and reducing financial pressure on the government. It is also conducive to enhancing the provision efficiency of public services by the government and improving the quality of public services by the government.

8.1.4 *The Need to Promote Social Development and Foster Social Organizations*

It is vital for social development to foster social organizations. By purchasing public services from social organizations, the government boosts social organizations and promotes social growth. Lessons abroad about government purchase of public services tell us that social organizations are important providers from which the government purchases public services. According to the research of the Johns Hopkins Comparative Non-Profit Sector Project, in 39 countries with reliable data, the average proportion of government funding in the income of social organizations is 36%. In 14 countries among them, the government is the largest income source of social organizations.[7] Government purchase of public services from social organizations provides stable economic support for social organizations and facilitates their growth and development. The practice is conducive to forming the "government-enterprise-society" ternary structure.

8.2 Government Purchase of Public Services in Policy Agenda

Luoshan Civil Community, jointly established by the Shanghai Bureau of Social Development, Shanghai Pudong New Area Social Development Foundation, and Shanghai YMCA in 1995, is regarded as the earliest non-profit organization in China from which the government purchases public services. It created a new model of public service provision, in which a Chinese local government purchased public services from non-profit organizations. Since then, local governments have sought and implemented purchase of public services. Governments of Beijing, Shanghai, Nanjing, Wuxi, Guangzhou, and Shenzhen have made more and more trials of purchase of public services from non-governmental organizations (NGOs) in different forms. The purchase extends to the areas of education, public health and AIDS prevention, poverty alleviation, elderly care, services for people with disabilities, community development, community correction, culture, urban planning, civic education, environmental protection, policy consultation, and many other aspects.[8] The Chinese government purchase of public services has achieved a new mode of government purchase of public services with Chinese characteristics, in which pilots drive the whole region and lessons from pilots are promote around the whole region.

[7] Wang Puqu, Lester M. Salamon, et al., *Outsourcing Government-Financed Social Services to Civil Society Organizations: Lessons from China and abroad* (Beijing: Peking University Press, 2010), 34.

[8] Jia Xijin, Su Ming, et al., *Final Report on Government Procurement of Public Services: People's Republic of China* (Asian Development Bank), 2009.

Government purchase of public services is being included on the policy agenda.

At the national level, the CPC Central Committee and the State Council have attached great importance to government purchase of services and released many arrangements and instructions. The 18th National Congress of the CPC emphasized the need to improve the way the government provides public services. The new government explicitly called for making more use of non-governmental players in public services and increasing government purchase of services. The Decision on Major Issues Concerning Comprehensively Deepening Reforms adopted at the Third Plenary Session of the 18th CPC Central Committee clearly stated to promote government purchase of services. For any administrative affair services, a competition mechanism should, in principle, be introduced. The services should be purchased from non-governmental players through contracts, commissions, and other means.[9] On July 31, 2013, Premier Li Keqiang presided over an executive meeting of the State Council to study and promote government purchase of public services from non-governmental players, to deploy and strengthen urban infrastructure construction. He made it clear that the public services that are suitable for market-oriented provision should be handed over to qualified and reputable social organizations, institutions, and enterprises. Later, the General Office of the State Council issued the Guiding Opinions of the General Office of the State Council on Effectively Implementing Government Purchase of Services from Social Forces. Relevant ministries and commissions have issued opinions on or measures for government purchase of public services in fields in their responsibilities in accordance with the essence of the CPC and the Central Government's instructions regarding the transformation of government functions and government purchase of public services. In 2015, 65 central departments carried out pilot programs for government purchase of services worth about CN¥9 billion in financial funds.[10]

At the local level, regions have developed regional and departmental implementation opinions, methods, and measures for government purchase of services according to the practice and exploration of the purchase of services, and defined the type, nature, and content of purchased services. Some regions have sorted out and listed the content and items of

[9] zhs.mof.gov.cn/zhengwuxinxi/zhengcefabu/201312/t20131209_1022116.html.
[10] http://news.xinhuanet.com/fortune/201401/02/c_125947032.htm.

purchased services based on the needs of the people, and developed a guiding catalog based on careful research for the local government to purchase services. By the end of 2015, 27 provinces (autonomous regions and municipalities directly under the Central Government) had issued a guideline for government purchase of services. In 2015, Beijing, Shandong, Jiangsu, and Anhui spent more than CN¥10 billion in government purchase of services. In general, the atmosphere and mechanism for the Central and local governments to jointly promote the reform of government purchase of services has taken shape.[11]

8.3 Problems in and Defects of Government Purchase of Public Services

Government purchase of public services has become the main form of government provision of public services in developed countries and regions. In China, however, government purchase of public services from social organizations is still something new, as most public services have long been provided by the government. As the existing model of public service provision by the government is turning to government purchase of public services from social organizations, both the government and the public find themselves unaccustomed in terms of systems and cultural psychology.[12] Donald Kettl poured cold water on privatization. He divided the defects that lead to market failure in contract outsourcing into "supply-side defects" and "demand-side defects." The former refers to the defects in the market in which the government provides goods and services, and the latter refers to the defects of the government itself as a buyer.[13] Based on Kettl's division, the risks or problems in the Chinese government purchase of public services from social organizations are analyzed from three perspectives: the supply side, the demand side, and beneficiaries.

[11] http://news.xinhuanet.com/fortune/201401/02/c_125947032.htm.

[12] Wang Puqu, Lester M. Salamon, et al., *Outsourcing Government-Financed Social Services to Civil Society Organizations: Lessons from China and abroad* (Beijing: Peking University Press, 2010), 27.

[13] Zheng Weidong, "A Study on the Government Purchase of Public Services in Urban Community Construction," *Journal of Yunnan University of Finance and Economics*, no. 1 (2011):153–160.

8.3.1 The Supply Side

The Overall Strength of Social Organizations Is Not Strong, and There Has Not Been a Sufficient Competitive Market

The NPM theory that flourished in the West in the 1980s is the theoretical foundation of government purchase of public services. The core idea of this theory is the market mechanism and full competition. Therefore, the effectiveness of government purchase of public services largely depends on the participation of sufficient and diverse social organizations in competition. In today's China, the number of social organizations is still inadequate compared with the country's huge population and insufficiency in public services. In addition, resources for social organizations from the government are very limited. There are just a few small-sized social organizations engaged in purchasing, and there has not been a sufficient competitive market.[14] The number of non-profit organizations registered with the civil affairs department in China is about 400,000. However, according to statistics, the actual number should be at least 3 million. For a long time, due to the pre-approval policy and cumbersome procedures of the civil affairs department, most NPOs have only existed as affiliations and grass-roots teams. Due to excessive administrative intervention, social organizations in China are dependent, insufficient, small in total number, understaffed, irrationally structured, and unevenly distributed regionally. It is difficult for these underdeveloped social organizations to undertake large-scale public service entrustment from the government. In 2013, the Government Procurement Center for the State Organs called for bids for home-based care services for seniors with disabilities and semi-disabilities by central state organs. Suppliers' feedback and experts' demonstrations showed that there was only 1 elderly care service institution able to serve 100 persons per day in the Beijing market. In the end, the center had to apply to the finance department for single-source procurement.

Social Organizations Are Weak and Lack Bargaining Power

Social organizations undertake and provide public services purchased by the government. Their development, to some extent, determines the

[14] Zheng Weidong, "A Study on the Government Purchase of Public Services in Urban Community Construction," *Journal of Yunnan University of Finance and Economics*, no. 1 (2011):153–160.

effect of government purchase of public services.[15] Only capable and eligible social organizations can complete service tasks with high quality. From the perspective of the origin, some Chinese social organizations are separated from government agencies or institutions, and some of them are even established after being entrusted some purchase tasks. They are more officially related and less independent. Other Chinese social organizations have developed freely in society. These organizations have complete internal structure and are staffed with less capable workers. The workforce has unreasonable educational, knowledge, and age distributions, and just a few of them are professionals and full-time workers. The organizations' operation is not sufficiently standardized, without sound service awareness or service capabilities.[16] Therefore, social organizations as the undertaker of purchase of public services, which are dependent or internally defective, and the government as the purchaser of public services are unequal in negotiating and constatation.

Social Organizations Lack Social Credibility and Are Not Recognized by Society and Service Recipients

Traditionally, the Chinese public has considerably trusted the government and been accustomed to the existing model of government provision and production of public services. When social organizations start to undertake public services in place of the government, the public generally do not trust them. Besides service recipients, government workers also do not trust or even resist social organizations. The government tends to be very prudent over purchasing public services from social organizations. The public's distrust of social organizations in undertaking public services increases costs of purchase of public services.[17]

[15] Yu Chongsheng and Xu Honglin, "Promoting Government Purchase of Public Services Is the General Trend," *People's Tribune*, no. 32 (2013): 6–11.

[16] Editorial of *China Social Organization*, "Local Government Purchase of Services from Social Organizations: Moving forward in exploration," *China Social Organization*, no. 10 (2013): 7–10.

[17] Wang Puqu, Lester M. Salamon, et al., *Outsourcing Government-Financed Social Services to Civil Society Organizations: Lessons from China and abroad* (Beijing: Peking University Press, 2010), 31.

8.3.2 The Demand Side

Lagging Transformation of the Government and Service Awareness
A necessary premise for government purchase of public services is that the government separates the "producer" and "provider" functions of public services. At present, in public service undertakings, the government often performs both the functions of provision and production. The omnipotence in providing public services greatly increases the costs of public service provision and reduces the efficiency and benefit of government purchase of public services. Therefore, as of government purchase of public services, it is necessary to center services on citizens and society. Government functions should be comprehensively transformed from regulation to service, from omnipotence to limitedness, from economic development to public services. A service-oriented government should be created.[18]

Unclear Purchase Boundaries
In the process of government purchase of public services, there is a lack of clear standards and specifications for which services should be purchased and which must be undertaken by the government. The Guiding Opinions of the General Office of the State Council on Effectively Implementing Government Purchase of Services from Social Forces notes that the services purchased by the government from non-governmental players should be suitable for market provision and within the capability of non-governmental players. However, in practice, there is a lack of clear definitions for which public services are suitable for market provision and which fall in the capability of non-governmental players.

Narrow Purchase Areas
At present, government purchase of public services focuses on general social areas, including home-based elderly care, migrant worker training, social welfare and assistance, disability rehabilitation, social correction, drug control, and marriage and family. It is a rare practice in leading public service areas such as public education, scientific research, medical care, and social security.

[18] Yu Chongsheng and Xu Honglin, "Promoting Government Purchase of Public Services Is the General Trend," *People's Tribune*, no. 32 (2013): 6–11.

Single Purchase Model

According to whether the social organization undertaking public services is independent from the purchasing government department and whether the purchase is competitive, Chinese social organizations' engagement in the purchase of public services can be divided into independent competitive purchase, independent non-competitive purchase, dependent competitive purchase, and dependent non-competitive purchase.[19] In the current practice of government purchase of public services in many parts in China, dependent non-competitive purchase dominates. In most cases of it, the government establishes non-profit organizations to undertake some of its services or administration functions and provides funds and resources in projects or other forms. Such purchase of public services is a one-dimensional cooperation with the government. Under this model, government departments purchase services mostly from organizations with administrative relations or other affiliation with them. In the name of undertaking public affairs entrusted by the government, these social organizations just accept tasks assigned by the government at any time. They are just extensions of government departments and "secondary governments" with the same administrative attribute as the government.[20]

Less Regulated Purchase Process

The Chinese government purchase of public services from social organizations is still in the pilot and exploratory stage, and government purchase of public services still lacks a national legal framework. The 2003 Government Procurement Law includes just the purchase of logistics services that maintain the government's operations. Public services closely related to people's livelihood are not included in the scope of procurement, which makes government purchase of public services lacking a legal and institutional framework. Moreover, government purchase of public services lacks standardized procedures, with the budget, purchase process, purchase results usually not disclosed. Consequently, it is either a

[19] Wang Ming and Le Yuan. "Analysis on the Models of Chinese Non-governmental Organizations' Involvement in Purchase of Public Services," *Journal of Zhejiang Provincial Party School*, no. 9 (2008): 5–13.

[20] Wang Puqu, Lester M. Salamon, et al., *Outsourcing Government-Financed Social Services to Civil Society Organizations: Lessons from China and abroad* (Beijing: Peking University Press, 2010), 28.

de-institutionalized process based on social connections or an essentially internalized cooperation.[21]

Absence of a Purchase Supervision and Evaluation System

The Chinese government purchase of public services lacks a sound market-oriented competition mechanism and effective supervision, management, and evaluation of quality in the process. On the one hand, it is difficult to quantify service recipients' evaluation of the effect of some "soft services" (such as education, elderly care, and personal services) and calculate and measure their costs and prices. On the other hand, cost accounting, price determination, and service quality standards are highly specialized, which results in the information asymmetry among purchasers, undertakers, and beneficiaries. It is impossible to supervise and evaluate professional technical indicators such as the use of funds, purchase prices, and service levels in purchase of public services by relying only on the internal supervision of government departments without an independent third-party regulatory mechanism. At the same time, as Chinese government purchase of public services is still in the initial stage of exploration and the system is not perfect, government purchase of public services is somewhat arbitrary. As a result, there is a large room for rent-seeking in purchase of public services, which leads to corruption and even conspiracy between the government and non-governmental organizations to infringe on state assets.

8.3.3 Beneficiaries

The public is the direct beneficiary of government purchase of public services. Good education and altruistic civic awareness are the social foundation for government purchase of public services from social organizations. Only when citizens understand and trust the public services provided by the government and social organizations will it be practical for the government to purchase public services from social organizations and provide them to citizens.[22] For now, the Chinese public still have weak civic awareness and need to enhance their citizenship.

[21] Wang Puqu, Lester M. Salamon, et al., *Outsourcing Government-Financed Social Services to Civil Society Organizations: Lessons from China and abroad* (Beijing: Peking University Press, 2010), 29.

[22] Wang Puqu, Lester M. Salamon, et al., *Outsourcing Government-Financed Social Services to Civil Society Organizations: Lessons from China and abroad* (Beijing: Peking University Press, 2010), 42.

8.4 Goals, Areas, and Approaches of Government Purchase of Public Services in 2020–2030

The period from 2020 to 2030 will be a critical period not only for comprehensively building "a moderately prosperous society" but also for accelerating the making of China into a modern socialist country that is prosperous, strong, democratic, culturally advanced, harmonious, and beautiful. China in 2030 will be a great modern a country with a stronger economy, greater democracy, greater social harmony, thriving culture, and ecological balance. At that time, China's economic and social development indicators will basically reach the level of moderately developed countries. Some researchers have made the following predictions:[23]

In terms of economic aggregate, in 2020, China remained as the world's second largest economy, with a per capita GDP up to US$7500, reaching the level of moderately wealthy countries and close to the level of moderately developed countries. In 2030, China is very likely to surpass the United States and become the largest economy in the world, with a per capita GDP of US$12,500, basically reaching the level of moderately developed countries and becoming one of the middle- and high-income countries.

In terms of industrial structure, in 2020, the proportion of the tertiary industry exceeded 50%, surpassing the primary and secondary industries to become the leading industry. In 2030, the tertiary industry will absolutely dominate in China, accounting for 62.37%. China will transform from an economy dominated by industry to one by services, especially modern services.

In terms of urbanization, in 2020, China's urban population accounted for more than 60% of the total. In 2030, this figure will increase to 68.9% to 71.7%, possibly approaching or surpass 72%, basically reaching the level of moderately developed countries.

In terms of population size and structure, in 2020, China's population exceeded 1.4 billion, and those aged 65 and above increased to 13% of the total. In 2030, China's population will peak at about 1.5 billion, and those

[23] Forecast data of China's economic aggregate, industrial structure, urbanization rate, and population structure from 2020 to 2030 and its basis can be found in the *Report on the Modernization of Public Services in 2030: Analysis and Forecast of the Macro Background for China's Public Service Development in 2030* by Liu Peng.

aged 65 and above will surpass that of Japan, making China the country with the largest aged population in the world.

In terms of fiscal revenue, in 2020, the fiscal revenue of Chinese governments at all levels was about CN¥24,493.06 billion, and the fiscal expenditure was about CN¥25,760.05 billion. In 2030, the fiscal revenue of Chinese governments at all levels was about CN¥63,528.17 billion, and the fiscal expenditure will be about CN¥75,692.86 billion.

In terms of the development of social organizations, in 2020, the total number of social organizations exceeded 800,000, and the number of professionals at social organizations exceeded 10 million. In 2030, the total number of registered social organizations in China will be 1.693 million, including 1.042 million social groups, 648,000 private non-enterprise entities, and 57,000 foundations. The number of professionals at social organizations will exceed 20 million.

8.4.1 *Goals*

Toward the strategic goal of building a moderately prosperous society in an all-round way, taking into account Chinese government purchase of public services at present and China's economic and social development in 2020, the main goals of Chinese government purchase of public services in 2020 are as follows.

First, terms of reference for guidance will be set. A nationwide and complete system of government purchase of public services from non-governmental players will be basically established. A unified and effective platform and mechanism for purchase of services will be basically formed.

Second, pilot programs will drive comprehensive roll-out. Government purchase of services from non-governmental players will be fully implemented everywhere. A more reasonable public service resource allocation system and supply system that is compatible with economic and social development will be formed.

Third, supply will be increased and methods will be innovated. Government investment will increase substantially, and the proportion of budgetary expenditure for basic public services in fiscal expenditures will also increase. There will be innovative public service supply methods, and providers and methods of delivery will be significantly more diverse.

Fourth, the development will be balanced and the general outcome will be satisfactory. The gap between urban and rural areas in basic public services will be significantly narrowed, and the equalization of basic public services will be generally achieved. An effective mechanism for urban and rural residents to express their needs for basic public services will be established. There will be sound systems for performance evaluation and administrative accountability. Social satisfaction will be improved.

Toward the great goal to make China a modern socialist country that is prosperous, strong, democratic, culturally advanced, harmonious, and beautiful, as China's economic and social development approaches and reaches the level of moderately developed countries or newly industrialized countries, the main goals of Chinese government purchase of public services in 2030 are as follows.

First, systems will be complete and mechanisms will be optimized. A sound system for government purchase of public services from non-governmental players will be built. A systematic, holistic, and coordinated institutional system for government purchase of public services will be formed. The public service provision mechanism will be optimized. A new public service provision mechanism will be established which is "diverse around one core, competitive for the optimal, open and transparent, strictly supervised, and adjusted from time to time."

Second, the guiding philosophy will be advanced and the boundaries will be clear. Non-governmental players will be promoted to deliver more and better public services to meet the diverse public service needs of the general public. In the field of basic public services, the government will gradually purchase more services from non-governmental players. In the field of non-basic public services, non-governmental players will play a more important and better role. For any administrative affair services, a competition mechanism should, in principle, be introduced. They should be purchased from the private sector through contracts, commissions, and other means.

Third, it will be equitable and efficient and satisfy the people. In 2030, basic public services will be basically equalized. Goods and services in basic public service areas such as education, employment, social security, health care, housing security, culture and sports, and services for disability will be guaranteed to be efficiently provided. The people will be basically satisfied with the public services provided by the government.

8.4.2 Areas

The areas of government purchase of public services define the boundaries. With regard to the boundaries of government purchase of public services, Robert Mercer believed that, as far as local governments are concerned, every service it provides and every function it performs can be outsourced. However, it is more generally recognized that not all services are suitable for outsourcing, and there should be certain criteria. E. S. Savas pointed out in his *Privatization and Public-Private Partnerships* that outsourcing public services originally undertaken by the government to private enterprises through contracts makes the government a purchaser of public services and private enterprises providers of public services. The effective performance of such contracts requires some conditions. First, the tasks can be clearly defined. Second, there are several potential competitors. Third, the government can monitor the performance of the contractor. Fourth, the contracting conditions and requirements are prescribed in the contract and guaranteed to be implemented.

To define the boundaries of government purchase of public services, we must first classify public services. According to public goods' exclusivity and competition in consumption, we can divide public services into basic public service areas and non-basic public service areas. The former fall within the purview of the government as clearly stipulated in the Constitution. Although they are "non-exclusive" as a kind of public goods, there are distinctions between selectivity and non-selectivity, competitiveness and non-competitiveness in its consumption. The latter are not within the government's responsibilities expressly stipulated in the Constitution and without exclusivity but there are multiple competitors. In nature, they are similar to "private goods." According to the above division, the boundaries and areas of government purchase of public services can be defined as follows:

The first is public security, national defense and security, foreign affairs, civil affairs, judicial administration, and supervision. They are non-selective and non-competitive basic public service areas as expressly stipulated in the Constitution and fall within the government's responsibilities. As they are not exclusive and there are no other competitors, these public services should be produced by the government but not by non-governmental players. The government should directly provide public services in these areas instead of purchasing them from non-governmental players, as these services are prohibited from outsourcing.

The second is services closely related to people's livelihood (education, employment, social security, medical and health care, housing security, culture and sports, urban public transportation, etc.), public facility management services (urban planning and design, municipal public facility management, landscaping, urban appearance management, park services, etc.) and environmental services (urban sanitation, the control of water, air, noise, and waste pollution, etc.). None of them are exclusive but there are competitors in the market offering similar services. Therefore, they belong to the field of basic public services that are selective and competitive, and the government can provide such basic public services by purchasing from non-governmental players.

The third is professional technical services, including scientific research, experiment, and development; market research and public opinion polling services; social and management consulting services; employment services; technical testing and appraisal services; earthquake monitoring and forecasting services; meteorological observation and forecasting services; and surveying and mapping services. They all fall in the government's responsibilities or belong to the non-basic public service areas that help the government to better perform its duties. They are somewhat competitive services. Therefore, in order to give more and better play to non-governmental players, the government can provide these services by purchasing.

The fourth is convention and exhibition services; accommodation and catering services; engineering consulting and management services; real estate services; energy production and distribution services; financial services; transportation services; agriculture, forestry, animal husbandry, agricultural byproduct, and fishing services; mining and manufacturing services; and service items such as wholesale and retail services. They are beyond the purview of government responsibilities and in the area of general services as "private goods." They are exclusive and have multiple competitors at the same time. Therefore, these services can be provided completely by market players and through market mechanisms such as enterprises.

8.4.3 Approaches

Since the 1980s, the marketization of public services has been the main direction of public service reforms in Western countries. The forms of marketization mainly include contract lease, public-private partnerships, user payment, franchising, issuance of bonds, and the securities system (or voucher system). Among them, contract lease is common. In the United States, contract lease accounts for almost half of marketized public services. At present, more than 200 US government services are provided by contractors.[24] It can be said that in Western countries, contract outsourcing is the most common way of marketization of public services. When promoting government procurement of public services, China can learn from the experience and methods of contract outsourcing in Western countries, and promote government procurement of public services through various cooperation channels such as contract outsourcing, project application, and direct funding.

In most cases of contracting outsourcing, the government signs a service contract with a private entity for a service and pays the fees according to the contract. The non-governmental player undertakes and produces the specified public service item.

In most cases of project application, the government, as the purchaser of public services, designs special projects according to its needs and invites social organizations to bid. Social organizations, according to their needs, apply for the project to relevant government departments. The government reviews the application and provides financial support in the project. The social organization provides services according to project requirements.

Direct funding refers to the government granting funds to social organizations undertaking public services through financial subsidy, in-kind subsidy, or preferential policy support. The government can also use the warrants or consumer vouchers, that is, issuing vouchers to eligible public service recipients. Consumers of public services choose service providers and deliver the vouchers to them. With the vouchers, the service provider asks the competent authority to cash in a certain amount of funds.

8.4.4 Supervision

In order to ensure that non-governmental players' provision of public services is qualified and achieves the established goals, it is necessary to establish a comprehensive evaluation system with institutional arrangements for service quality to ensure that non-governmental players' provision of public services operates in an orderly and healthy manner.

First, an independent third-party evaluation mechanism should be introduced. The independence and professionalism of third-party agencies

[24] Yang Liqiong, "The Comparison of Public Service Marketization between China and America and its Implications," *Chinese Public Administration*, no. 7 (2011): 66–69.

should be fully exerted. The comprehensive performance of services entrusted to social organizations should be monitored and evaluated to timely identify problems and make improvements, so as to ensure that the services entrusted by the government to social organizations can achieve the set goals.

Second, a comprehensive evaluation system involving diverse parties should be formed. A sound supervision mechanism requires to develop a diversified evaluation system composed of independent third-party evaluation agencies, government departments, media, experts and researchers, the public, and service recipients, so as to increase the perspectives and breadth of the evaluation.

Third, a dynamic evaluation system for the whole process should be established. Dynamic supervision departments and evaluation measures should be in place for steps of non-governmental players' participation in public service projects, including application, review, project approval, bidding, contract signing, implementation, project conclusion, evaluation, and feedback. It is for tracking and managing the service process. Problems that arise should be studied, with feedback provided and improvements made in a timely manner.

Fourth, a comprehensive evaluation system should be developed. Before a project starts, it is necessary to determine the basic public services to be undertaken by non-governmental players and the conditions for qualified social organizations based on demand and organizational qualification evaluation. In the course of a project, the focus should be placed on process evaluation, which should be carried out in various ways such as random inspections by government departments, media supervision, public feedback, and self-discipline evaluation of the service organizations to ensure that the social organizations perform the agreement. After a project is concluded, the focus should be placed on the evaluation of government entities, so that the people's view of public services is truly reflected in the evaluation and appraisal of relevant government departments and officials. It demonstrates the ultimate responsibility of the government. At the same time, service quality should also be evaluated, including the evaluation of the economy, efficiency, and effectiveness of the public services, and the satisfaction of the public services, and their convenience.

8.5 International Experience of Government Purchase of Public Services

Government purchase of public services is one of the ways of public service provision. It is an important change made by Western governments to their social welfare systems. Public services emphasize that the government should guard citizens' basic survival rights and achieve social equity. Globally, the practice and reform of government purchase of social public services are usually associated with the privatization reform of the welfare state. As the Western philosophy on public services and social governance further develops, public service provision has undergone a process with equal emphasis on market-oriented reform and socialized reform.

8.5.1 Legislation First Becomes an International Common Practice of Government Purchase of Services

Almost all countries that carry out purchase of public services have relevant legislation at the national level. Most of them have relevant local legislation. For example, in 1998, the then British Prime Minister Tony Blair signed the Compact on Relations Between Government and the Voluntary and Community Sector in England (the Compact). After that, local governments developed local compacts based on the Compact. The Compact defines the partnership between the government and the voluntary sector and sets an overarching framework for public services to facilitate effective collaboration between the government and the community.

In 2012, after long efforts, the Public Services (Social Value) Act was unanimously approved by the British House of Lords, which means that the UK officially regulates government purchase of public services in law. It requires social organizations, when performing public service contracts, to consider how to improve the economic, social, and environmental well-being of the community they serve.

In 2006, in order to standardize the operation of government purchase of services, the Japanese government introduced the Law on Introducing a Competition Mechanism to Reform Public Services. The law sets the basic guidelines for public service reform and defines the goals and significance of the reform. It regulates the procedures of public-private bidding and private bidding, including the qualifications, invitation, and determination of bidders. It set the necessary measures for the bid-winning private entity to provide public services, including the signing and termination of contracts and supervision in the provision process. The law provides guidelines for government purchase of public services and a solid legal basis for the Japanese government's purchase of public services.[25]

8.5.2 There Are Various Forms of Participation, and Public-Private Partnerships Have Become the Mainstream

Since the 1980s, public service reform in Western countries have mostly adopted contract outsourcing, public-private partnerships, user payment, direct funding, and other cooperative approaches to promote government purchase of public services. The approaches include contract leasing, public-private partnerships, user payment, franchising, bond issuance, and securities (or warrants). Affected by their own history, culture, institution, and other factors, countries have made different choices in marketization approaches and methods.

One way is from privatization to public-private partnerships. After the Second World War, the UK rolled out large-scale nationalization and a welfare system, and established a "welfare state" in which the government provided comprehensive public services "from the cradle to the grave." Most of the public services related to the national economy and people's livelihood, such as housing, transportation, education, medical care, and unemployment assistance, were provided by the government at low prices. Most monopoly industries such as energy, electricity, postal services, and railways were also controlled by the government. The early efforts to reform the public service provision model in the UK were mostly changing the role of the government in public undertakings, vigorously promoting privatization plans, and advocating "reduction of state-owned borders." Former British Prime Minister Margaret Thatcher proposed measures to promote privatization in the 1979 Conservative Party Manifesto. The major practice was to sell state-owned real estate and property and some state-owned enterprises and put them on the market.[26] Following Margaret Thatcher's reform, Tony Blair proposed the Third Way during his tenure as British Prime Minister and continued to promote reform. This proposition requires identify areas of public services for the

[25] Han Lirong, Sheng Jin, and Gao Yubin, "A Review of System of Japan's Government Purchase of Public Services," *Contemporary Economy of Japan*, no. 2 (2013):15–21.

[26] Yao Jun, "Outsourcing Contract of Public Service in the U.K.: The History and Policy Development," *Science and Technology Management Research*, no. 14 (2014):192–197.

government and the market to provide and for public-private partnerships. Driven by the Third Way, the UK has made an attempt to provide public services in partnerships between the public and the private sectors.

The other is from contract outsourcing to public-private partnerships. Unlike the UK and other Western countries, the United States has taken contract outsourcing of public services or government service in the marketization of utilities. In service outsourcing, private enterprises and some social organizations can provide a variety of public services, including park management, health care, preschool education, social housing, elderly care, and community justice correction services. According to E. S. Savas' calculations, there are at least 200 services provided by contractors to the government in the United States. By 1987, 99% of towns with a population of more than 5000 and counties with a population of more than 25,000 had carried out contract outsourcing. From 1982 to 1997, local governments used contract outsourcing for 90% of their services when choosing alternative methods for service provision.[27] After 1997, the proportion of public-private partnerships in the provision of all services has been increasing. It was 18% in 1992, 17% in 1997, and 24% in 2002.

8.5.3 Government Purchase of Public Services Covers Most Public Service Areas, and Contract Outsourcing Projects Have Become the Core Functions of the Government

Government purchase of public services covers most public service areas, mainly including: elementary, vocational, and special education; public health; elderly care services; disability services; community services; cultural and traditional preservation; addressing employment, affordable housing, and other social issues; and consultations on urban planning, transportation, environment, and other policies. In the United States, contract outsourcing has penetrated into the core functions of the government. Even prisons, seen as symbols of the state apparatus, have become an industry in the country, as the federal government and 31 states have signed escrow agreements with private prisons. In 1991, a review by the Federal Audit Office of outsourcing contracts of the Department of Transportation, the Department of Energy, and the Environmental Protection Agency showed that in 28 of 108 contracts randomly selected,

[27] Ju Hua, "Contracting Out Public Services in U.S. Local Governments: Trends and inspiration," *Chinese Public Administration*, no. 7 (2008): 103–107.

the contractors performed tasks that were essentially functions that should be performed by the government as prescribed by laws and policies, including writing testimony, deciding whether to conduct loyalty investigations, and developing policies.[28]

8.5.4 *Non-profit Organizations Are Important Undertakers for Government Purchase of Public Services*

Various types of institutions in developed Western countries can be potential providers of social public services, including registered charities, non-profit organizations, associations, self-help groups, community groups, and social enterprises. Both in the privatization model represented by the UK and in the service outsourcing model represented by the United States, , NGOs are regarded as important partners. Some countries and regions regard non-profit organizations as the primary force for social welfare. Research shows that 40% to 70% of the income of most non-profit organizations in Europe comes from public financial resources. The figure is 45% in Japan and 70% to 80% in Hong Kong, China. Even in a market-dominated country like the United States, public sector resources still account for about 31% of the total income of NPOs. According to statistics, the British government spends up to GB£236 billion in purchase of public services every year. About 11% of these public service contracts are executed by social enterprises and charities. It brings high cost-benefit services to the government and supports the development of social organizations. In addition, the British government distributes 16.7% of the annual gaming revenue to charities across the country through government funds and set up special Treasury funds for public welfare undertakings.

8.5.5 *Open Bidding Is the Most Typical Way for Purchase*

Open bidding is the most typical way for purchase. The 1990 National Health Service and Community Care Act of the UK clearly stipulates that 85% of the special funds appropriated by the Central Government must be spent on services with competitive bidding from private or non-governmental organizations. According to the Law on Introducing a

[28] Ju Hua, "Range of Applications of Contracting Out for Public Services: Contrast between theory and practice," *Chinese Public Administration*, no. 4 (2010): 51–55.

Competition Mechanism to Reform Public Services, the Japanese government's public service reform has promoted almost all government public service projects to be outsourced to professional companies through competitive bidding, making it the largest public sector reform in the Japanese history.

8.5.6 *Specialized Regulatory Agencies Are Established, and the Focus of Supervision Shifts from Process to Outcome*

The Compact on Relations between Government and the Voluntary and Community Sector in England has a commission headed by a commissioner. It is the independent body for overseeing the performance of the compact. The National Council for Voluntary Organizations participates in the management as a third party, who supervises any breach of agreement by either party and makes recommendations for possible obstacles and future adjustments in the government's purchase or non-profit organizations' provision of services. Moreover, the Charity Commission for England and Wales releases recommendation announcements for the government to effectively monitor and evaluate services, creating a good social image of a fair and just government for the benefit of the public. Japan's Law on Introducing a Competition Mechanism to Reform Public Services stipulates the establishment of a third-party agency, the Public and Private Bidding Supervision Committee, to ensure the transparent, impartial, and fair implementation of the public service reform.

Process supervision and outcome supervision lead to two different approaches to supervision. In the United States, the difference between the two mainly lies in two forms of contracts: design contracts and performance contracts. In a design contract, the government sets standards for the services and has the compelling responsibility for the standards. Design contracts are characterized by paying for input and process. The standards set by the government constrain the conduct of the social organization, which bears less responsibility. In a performance contract, payments are made based on the performance of the service provider. Performance evaluation promotes the continuous improvement of service standards and save more costs every year. Performance contracts are characterized by paying for results, and social organizations are highly responsible for their performance.

8.5.7 Voluntary Mechanisms Are Introduced, Emphasis Is Put on Civic Engagement in Public Services

By introducing voluntary mechanisms, more volunteer service resources are mobilized and put into public service provision. It includes vigorously developing volunteer services; promoting institutionalized, standardized, and legal volunteer services; and facilitating the complementary and organic integration of volunteer services and government services. It also includes actively developing charitable causes to enhance the charitable awareness of the whole society; actively fostering charitable organizations; and improving laws, regulations, and tax relief policies for charitable donations. These initiatives can give full play to charity in the provision and financing of basic public services.

Public services in Germany are characterized by the emphasis on civic engagement. Citizens have a direct vote on public services can participate in projects at different implementation stages. Germany also encourages vulnerable groups to become volunteers and engage in public service provision.

The Cameron government of the UK has launched the Big Society program. Communities, social organizations, non-profit social enterprises, and individuals have also started to provide services to each other. In particular, community residents act as designers, providers, producers, and users in community operations. They can design public services according to individual needs. A social network integrating democratic design, decision-making, implementation, and supervision will form in the community. All community residents together prepare local development plans, self-manage the community, and provide services by themselves.

References

Cai Fuchao, "Let culture reach everyone: The development of China's radio, movie, and TV public service system," *Qiushi*, no. 17 (2012): 22–24.

Cheng Zhen, "The Chinese government's publications and government information dissemination," *Journal of The National Library of China*, no. 4 (2001): 23–29.

China Center for Modernization Research, CAS, "China's national strength will surpass Japan and become the second largest in the world by 2030," *Northeast Electric Power Technology*, No. 3 (2008): 8.

Editorial of *China Social Organization*, "Local Government Purchase of Services from Social Organizations: Moving forward in exploration," *China Social Organization*, no. 10 (2013): 7–10.

Fang Hanqi, *General History of Chinese Journalism Vol. 3* (Beijing: China Renmin University Press, 2004): 31.

General Research Team of China on Strategic Response to Ageing Population, *General Report of China on Strategic Response to Ageing Population* (Beijing: Hualing Publishing House, 2014): 191–192.

Guo Qingwang, Liu Xiaolu, Jia Junxue, "State financial capabilities and basic public education service system construction in the national 12th Five-Year Plan," *Renmin University of China Education Journal*, 01 (2011): 35–47.

H. Frederickson, *The Spirit of Public Administration*, trans. Zhang Chengfu, Liu Xia, Zhang, and Meng Qingcun (Beijing, China Renmin University Press, 2013), 54.

K. Dong, N. Wei, *Transforming China's Public Services*,
https://doi.org/10.1007/978-981-99-3943-5

Han Lirong, Sheng Jin, and Gao Yubin, "A Review of System of Japan's Government Purchase of Public Services," *Contemporary Economy of Japan*, no. 2 (2013):15–21.

Hu Angang, Yan Yilong, and Wei Xing, *China in 2030: Towards Common Prosperity* (Beijing: Renmin University of China Press, 2011), 62.

Jia Xijin, Su Ming, et al., *Final Report on Government Procurement of Public Services: People's Republic of China* (Asian Development Bank), 2009.

Jiang Yungen, "The outcome and enlightenment of public education service reform in other countries," in *China Education Policy Review*, ed. Yuan Zhenguo (Beijing: Educational Science Publishing House, 2011).

Ju Hua, "Contracting Out Public Services in U.S. Local Governments: Trends and inspiration," *Chinese Public Administration*, no. 7 (2008): 103–107.

Ju Hua, "Range of Applications of Contracting Out for Public Services: Contrast between theory and practice," *Chinese Public Administration*, no. 4 (2010): 51–55.

Li Shantong, Liu Yunzhong, et al, *China's Economy in 2030* (Beijing: Economic Science Press, 2011), 37. Hu Angang, Yan Yilong, and Wei Xing, *China in 2030: Towards Common Prosperity* (Beijing: Renmin University of China Press, 2011), 62.

Lin Yifu, "China's GDP to rank first in the world by 2030," *Brand*, No. 8 (2006): 83.

Liu Peng, Forecast data of China's economic aggregate, industrial structure, urbanization rate, and population structure from 2020 to 2030 and its basis can be found in the *Report on the Modernization of Public Services in 2030: Analysis and Forecast of the Macro Background for China's Public Service Development in 2030*; Liu Qingyuan and Wen Yingna, "Institutional Demands in the Government Purchase of Social Work Services," *Journal of Social Work*, no. 11 (2007):11.

National Bureau of Statistics of the People's Republic of China, *China Statistical Yearbook 1996* (Beijing: China Statistics Press, 1996), 69.

National Bureau of Statistics of the People's Republic of China, *China Statistical Yearbook 2014* (Beijing: China Statistics Press, 2014), 158.

National Bureau of Statistics of the People's Republic of China, *China Statistical Yearbook 2015* (Beijing: China Statistics Press, 2015), 21,33,35,62,214,856

Su Ming, Jia Xijin, Sun Jie, and Han Junkui, "Chinese Government Purchase of Public Services," *Public Finance Research*, no. 1 (2010): 9–17.

The Ministry of Civil Affairs of the People's Republic of China, *China Civil Affairs' Statistical Yearbook 2011* (Beijing: China Statistics Press, 2011): 112.

Wan Guanghua, "2030: China's urbanization rate to reach 80%," *International Economic Review*, No. 6 (2011): 99–111.

Wang Ming and Le Yuan. "Analysis on the Models of Chinese Non-governmental Organizations' Involvement in Purchase of Public Services," *Journal of Zhejiang Provincial Party School*, no. 9 (2008): 5–13.

Wang Puqu, Lester M. Salamon, et al., *Outsourcing Government-Financed Social Services to Civil Society Organizations: Lessons from China and abroad* (Beijing: Peking University Press, 2010), 27,28,34.

Wang Su, Zou Junwei, Sun Yuze, Fang Yong, "International comparative analysis of public education services," in *China Education Policy Review*, ed. Yuan Zhenguo (Beijing: Educational Science Publishing House, 2011).

World Bank and Development Research Center of the State Council, the People's Republic of China, *China 2030: Building a Modern, Harmonious, and Creative High-Income Society* (Beijing: China Financial and Economic Publishing House, 2013).65.

Yao Jun, "Outsourcing Contract of Public Service in the U.K.: The History and Policy Development," *Science and Technology Management Research*, no. 14 (2014):192–197.

Yan Qingchun, *Program of Action for the Development of Undertakings for the Aged in the New Era: An interpretation of the Plan for Development of China's Undertakings for the Aged During the 12th Five-Year Plan* (Beijing: China Society Publishing House, 2012): 135.

Yang Liqiong, "The Comparison of Public Service Marketization between China and America and its Implications," *Chinese Public Administration*, no. 7 (2011): 66–69.

Yu Chongsheng and Xu Honglin, "Promoting Government Purchase of Public Services Is the General Trend," *People's Tribune*, no. 32 (2013): 6–11.

Yu Guo'an, "Analysis and Policy Suggestions on Government Purchase of Public Services," *Review of Economic Research*, no. 46 (2011): 36–41.

Zhao Hongbin, "Educational competitiveness is the cornerstone of national competitiveness," *Education Science*, 4 (2008): 710.

FSC
www.fsc.org
MIX
Papier | Fördert
gute Waldnutzung
FSC® C083411

Kolibri 360 GmbH
Ferdinand-Jühlke-Straße 7
99095 Erfurt, Deutschland
produktsicherheit@kolibri360.de